# CAR AND DRIVER
# ON PORSCHE
## 1982–1986

Reprinted From
Car and Driver Magazine

ISBN 0 948207 82 5

Published By
Brooklands Books with permission of Car and Driver

# CAR AND DRIVER




*Distributed By*

Car and Driver
3460 Wilshire Blvd,
Los Angeles,
California 90010

Brooklands Book Distribution Ltd.
Holmerise, Seven Hills Road,
Cobham, Surrey KT11 1ES,
England

# CAR AND DRIVER

# CAR AND DRIVER

Car and Driver is a living, breathing entity.  Although some of our older editions may be a bit cranky and occasionally stiff-legged, they do live on for an eternity (if only on our shelves).  We do love to share our previous work, but it's obviously not possible to crank up the printing presses on a whim.  Instead, we've produced this series of books, each of which encompasses virtually everything said about a particular subject during a given period in Car and Driver.

We hope you enjoy these collections.  They have not been edited or updated in any way, so this is vintage Car and Driver at its finest (we think).



Printed in Hong Kong

# Porsche 944

## *The most seductive combination of economy and performance money can buy.*

The Porsche 944 is a great car. It renews your faith in Porsche, because the men who made the 944 put into it everything they know about fast cars. It renews your faith in fast cars, fast living, and fast women, because it brings you these things at a reasonable price and with fuel efficiency besides. From now on, Porsche performance is no longer restricted to those entertainment lawyers in L.A. who scuttle through Coldwater Canyon in their 911s and 928s on their way from their homes in Studio City to their warrens in Century City. The rest of us can finally have a piece of the action.

Of course, all Porsches are supposed to be serious automobiles. But sometimes it's been hard to keep the faith. First there was the 914, with its Volkswagen engine. Then came the 924, with its Audi drivetrain. Neither car was intended to be a Porsche: the 914 was marketed as a VW-Porsche in Europe and the 924 was developed to be an Audi sports car. Finally, the succession of 924s equipped

with tape stripes and special equipment hasn't bolstered Porsche's reputation as a maker of serious cars, either.

With the 944, however, Porsche has at last made over the 924 in its own image. The forthcoming demise of the 924's economical but rough-running engine, shared with such lesser forms as the AMC Spirit and one VW truck, provided the opportunity. Even in 1976, when the 924 was introduced, Porsche engineers recognized the limitations of the 2.0-liter four-cylinder and developed two alternatives. One was a 3.5-liter, 90-degree V-6 derived from the 928 engine; but experimentation with a 2.8-liter V-6 developed by Peugeot, Renault, and Volvo revealed that such an engine would weigh too much, consume too much fuel, run roughly, and be difficult to install on the assembly line. So it proved to be the second alternative that is being built at Porsche and installed on Audi's 944 production line.

This all-aluminum, 2.5-liter in-line four-cylinder duplicates the 924 engine's weight and fuel economy, but produces 30 percent more power. Just as important, the 944 engine carries two Mitsubishi-style balance shafts to exorcise objectionable engine vibes. More anti-vibration science went into the engine's mounting arrangement as well. Special shock-absorbing mounts attach the engine to a new, vibration-resistant aluminum crossmember, and the crossmember itself is connected to the chassis at points that are relatively insensitive to vibration. Just to emphasize that this engine, derived from the 928's V-8, is designed to do serious business, there's a Porsche logotype cast prominently into the valve cover.

The 944's cockpit environment has also been improved to eliminate the dentist-drill resonances that plagued the 924. First, the thick sound insulation developed for the 924 Turbo is carried over into the 944. The rack-and-pinion steering is mounted in rubber to improve isolation from road shocks, and the tie rods feature rubber bits for the same reason. Finally, a rubber-damped clutch and new transmission mounts keep the drivetrain from buzzing like a Swedish sexual appliance. Porsche's testing reveals that seat-of-the-pants vibes in the 944 are less severe than those of other sports cars (although it's unclear *which* sports cars), while engine noise is less than that of a conventional four-cylinder below 2500 rpm and equivalent to a smooth-running V-6 or V-8 above that mark.

As long as the guys in the lab coats at Weissach had the 924 laid out on the operating table, they decided to transplant some muscle into the chassis too. So the four-wheel disc brakes optional on the 924 Turbo are standard on the 944. The track has been increased 2.3 inches in front and 3.1 inches in back. Big 215/60VR-15 tires ride on cast wheels with seven-inch rims (205/55VR-16 tires on forged wheels with seven-inch rims are optional). Finally, weight distribution fore and aft is almost exactly equal and the suspension calibration falls midway between the 924 Turbo's and the European GTS Carrera's.

Wrapped around this package is the bodywork of the Euro Carrera, with its odd but powerful mixture of curves from the styling studio and angles from the racetrack. The 944 has a larger cross-sectional area than the 924, but a better drag coefficient (0.35, versus the 924's 0.36). A special windshield molding to reduce turbulence and a longer rear spoiler account for some of this improvement, but the major contribution comes from the elastic urethane nose, which routes air under the bumper and into the radiator very efficiently.

When you run all this stuff through your memory banks, of course, you realize that the 944 simply represents another brew of Porsche special equipment. Even the engine simply produces the same power as the 924 Turbo's. So why should this car be better? The point is, the 944 incorporates all of Porsche's *good* stuff. Porsche engineers built the 944 to be the kind of car that Porsche engineers like to drive.

You can feel the dedication to speed simmering inside the 944 even when you're droning down the turnpike. The car lopes along at about the same rpm as the 924 Turbo—both cars are geared almost identically, with very tall fourth and fifth gears—and 62-mph cruising delivers over 30 mpg and tremendous range from the 16.4-gallon fuel tank. Yet, unlike any 924, this car feels calm and composed, almost like a 928. And when you stick your foot in it, the high-compression engine gets you rolling right away without downshifting.

To be sure, the cockpit is more intense on the freeway than your average Chevy Caprice, but the 944 feels like a water bed compared with a 924, simply because engine noise and vibration are imperceptible. And the new ventilation system, which routes fresh air to all ducts and features 928 pieces in the standard air conditioning, takes the edge off the heat transmitted by all that glass surrounding the cockpit. You'll still find yourself wincing over expansion joints, though. Porsche engineers admit that the Rabbit suspension in the front and the VW microbus suspension in the rear just don't provide large enough rubber bushings to build in much compliance without making the handling all rubbery as well. So even though new top bushings for the front MacPherson struts improve road isolation somewhat, you'd be well advised to avoid Third Avenue in Manhattan and the entire state of West Virginia.

**B**ut listen, you want to know what the 944 is like once you crank it up to warp speed, right? Well, it's terrific. You can drive like a hero without sweat popping out on your brow. The 944 is great because it responds crisply and decisively to every command, and it builds up to its limits in a perfectly linear fashion. You won't find killer understeer here. And you won't find any nervousness at the limit either. You will find yourself catching the rhythm of Ortega Highway or Angeles Crest or your local racer road like a hill-climb champion. Feel free to hunt 911s if you wish, because even though the 944 gives away 29 hp and 130 pounds to the superest Super Beetle, it can be driven at the limit effortlessly, while it takes some concentration to drive the 911 more than six corners in a row without making a mistake.

It's as if the 944 had been developed in long drives through the Alps, rather than in short sprints around Porsche's Weissach test track. The cockpit environment at speed is calm. You always feel as if you are traveling much slower than the new yellow numerals on the speedometer tell you—that is, if the needle hasn't run out of numerals to read altogether. The suspension soaks up each bump and then damps the ride motion immediately. The engine revs crisply, but there's more than 130 pounds-feet of torque between 2500 rpm and 5500 rpm, so there's not much need to seek redline. The brakes are plenty powerful, although there's too much boost (just as with every 924 since 1976) and the pedal rides a little high for really effortless heel-and-toe action. There's no need to race the 944 to make time over public roads—you just drive it quickly, that's all.

# Technical Highlights

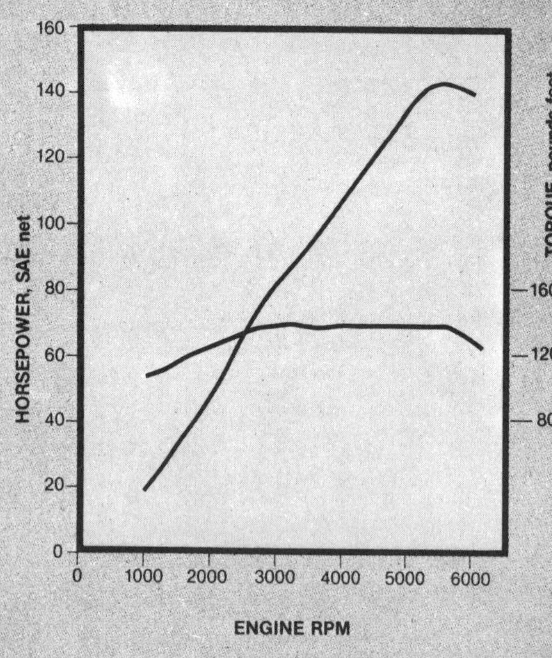



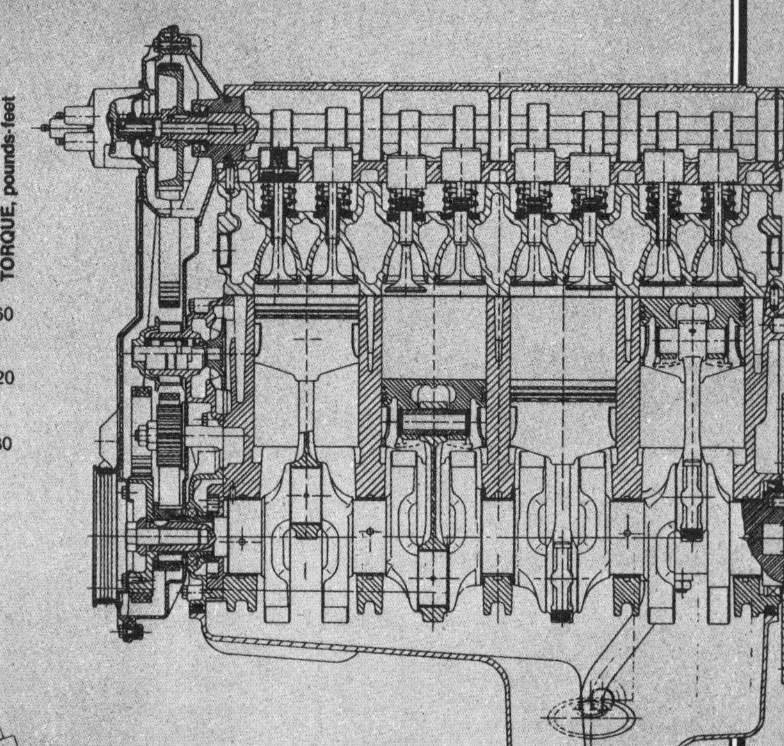

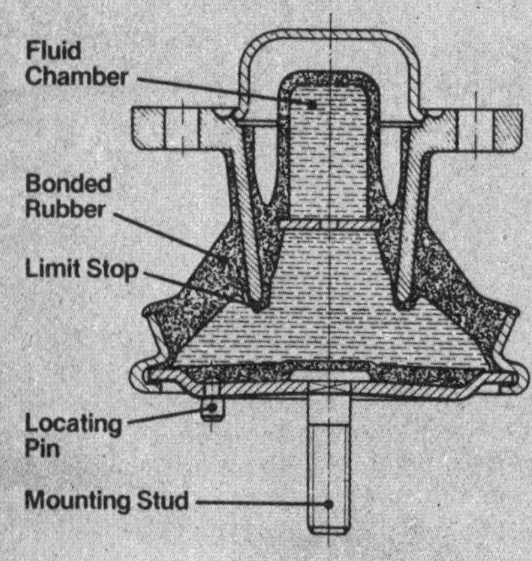



**Engine Mount**

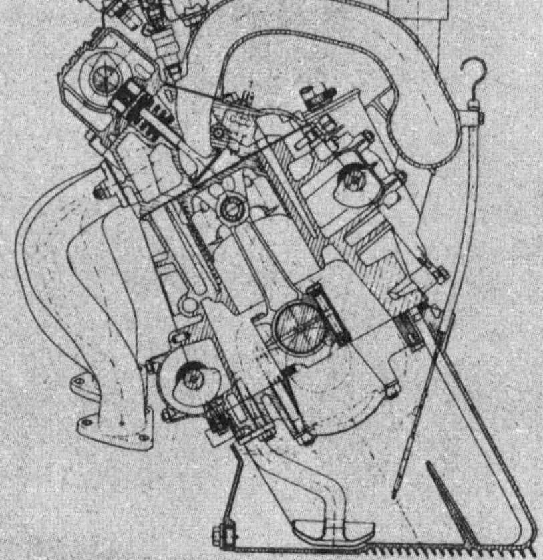

• The 924 is dead, after only seven years of production. Its demise, premature for a Porsche, was caused by its mediocre engine. The Audi-based powerplant was short on power and long on noise and vibration.

The 944 engine shows that these complaints eventually found sympathetic ears in Weissach. The new engine is basically one bank of a 928 V-8. Porsche's new four and its old eight share bore centers, cylinder-head-bolt spacing, Reynolds 390 die-cast-aluminum block construction (with linerless bores), aluminum cylinder heads with in-line valves and hydraulic lifters, sintered connecting rods, and a one-piece main-bearing saddle with an integrally cast oil gallery. Most of the actual parts are slightly different, but the castings are frequently the same, and they're machined with the same tooling.

Displacing 2479cc, the 944 engine is 25 percent larger than its 1984cc predecessor. Horsepower is increased more than proportionally, from 110 to 143 (a 30 percent improvement), and torque is up from 111 to 137 pounds-feet (a 23 percent improvement). The 944's 3000-rpm torque-peaking speed is 500 rpm lower than the 924's, and its torque curve is flat from 2500 to 5500 rpm.

Four-cylinder engines of this size normally shake like cement mixers. But Porsche engineers swallowed their pride in this case and purchased a license for twin balance shafts from Mitsubishi. The belt-driven shafts generate forces that cancel a four-cylinder's natural second-order (occurring at twice engine speed) vertical shaking forces. Positioned on either side of the block and vertically displaced, they also reduce the rocking movement inherent to four-cylinder engines. The balance shafts, originally patented by Frederick W. Lanchester of England in 1911, contribute eighteen pounds to the 330-pound engine weight and rob 4 hp at the peak, but they are still lighter and more efficient than the alternative of adding more cylinders.

Porsche has devised an innovative pair of engine mounts to take up where the balance shafts leave off in the struggle against roughness. Each mount is seated between an aluminum-alloy crossmember and a cast-aluminum engine bracket. Inside the rubber portion of each mount there are two antifreeze-filled cavities; a

And if you really want to press the limit, the 944 will not only let you, it will help you. The front end steers toward an apex easily—steering response is the one thing CN36 tires excel at—and as understeer builds up, the steering grows gradually heavier. At their low inflation pressures, the 944's front Pirellis squeal fairly easily, and if you enter a corner slowly and just turn the wheel you'll feel as if you're driving right over them. Enter a corner quickly, though, and it's another story. Trailing-throttle oversteer brings the rear end out just a touch to help point the front end toward the apex, and then the car's even weight distribution makes it easy to balance the 944 with gas pedal and steering wheel on the fast line. On the skidpad, the end result is 0.80 g, a respectable figure. But on the road, the result is magic, for there's one level of handling for everyday driving and another level for emergency maneuvers.

If you're looking for flaws in the 944, they do exist. The high bolsters of the leather sports seats fitted to this test car hold you in place, but they also emphasize the 944's lack of hiproom, make entry and exit from the driver's seat very difficult, and foul your elbows when you're working the steering wheel hard. The wheel still seems to sit in your lap,

no matter how small Porsche makes it. The Pirelli CN36 has been the rubber of choice for 924s since 1976, but it still rides too harshly even for this improved suspension. Furthermore, it isn't cheap. If you're looking for flaws, that's pretty much it. The rest is poetry.

Yet it's more than poetry that makes the 944 a great car. First, it's a great value. While "value" might be a difficult word to swallow in the same breath with a price just on the far side of $20,000, consider the facts. The price tag buys you a car with everything on it, from sunroof to air conditioning. All that's lacking is a radio to slot into the dash. The price tag also buys you a car that is comfortable on the Interstate and exquisitely exciting on a mountain road; there's no need to shore up the 944's performance with selections from the options list. Moreover, your money buys you a car that's as fuel-efficient on the freeway as a Rabbit. Suddenly the 944 becomes the bargain of the decade.

The thing that really makes the 944 a terrific car, though, is the conviction of the engineers who made it that only performance matters. Not market position, not price, just performance. And as so often happens, dedication to a simple ideal has improved both market position and value. The result is the best combination of performance and economy that money can buy—a serious car that advances the state of the automotive art.

After all, that's what we expect from Porsche.

—*Michael Jordan*

separator plate between the two volumes has an orifice that allows the liquid to flow from one chamber to the other as the engine moves about. This back-and-forth flow produces damping forces much like a conventional shock absorber. As a result, the rubber mounts can be soft for good isolation and still maintain proper engine location.

After great strides had been made in comfort, Porsche engineers concentrated on maximizing the 944's fuel economy. The combustion chamber appears at a glance to be a conventional wedge, but it benefits from Porsche's research into high compression with its TOP (thermally optimized performance) engine of a few years ago. The result is a fast-burn chamber that tolerates 9.5:1 compression even with America's low-octane unleaded fuels. Another fuel-economy plus is the Bosch Motronic engine-control system. This is Germany's version of the microprocessor engine controls common on American cars, although in the 944 it's coupled to port-type electronic fuel injection. It provides the same benefits of precise fuel metering and optimal spark timing with minimal maintenance.

Now that the low end of the Porsche line has been plumped up with new power, the opportunity for great gains at the high end has also arrived. A V-8 version of the new 944 engine could theoretically produce 300 net horsepower and 280 pounds-feet of torque in emissions-legal trim. It could power a real 928S into the hearts and minds of car enthusiasts across the land. Such a 928, with the same sort of revitalization the 924 has just received, would be the fastest automobile available in America.

—*Csaba Csere*

**Vehicle type:** front-engine, rear-wheel-drive, 2+2-passenger, 3-door coupe

**Price as tested:** $21,000 (estimated)

**Options on test car:** leather sport seats, digital AM/FM-stereo radio/cassette.

**Sound system:** AM/FM-stereo radio/cassette, 3 speakers

### ENGINE
Type . . . . . . . . . . . . . . . . . 4-in-line, aluminum block and head
Bore x stroke . . . . . . . . . 3.94 x 3.11 in, 100.0 x 78.9mm
Displacement . . . . . . . . . . . . . . . . 151 cu in, 2479cc
Compression ratio . . . . . . . . . . . . . . . . . . . . . . . .9.5:1
Fuel system . . . . . . . . . . . . . Bosch L-Jetronic fuel injection
Emissions controls . . . . . 3-way catalytic converter, feedback
                        fuel-air-ratio control
Valve gear . . . . . belt-driven single overhead cam, hydraulic
                                       lifters
Power (SAE net) . . . . . . . . . . . . . . . 143 bhp @ 5500 rpm
Torque (SAE net) . . . . . . . . . . . 137 lbs-ft @ 3000 rpm
Redline . . . . . . . . . . . . . . . . . . . . . . . . . . . . . 6400 rpm

### DRIVETRAIN
Transmission . . . . . . . . . . . . . . . . . . . . . . . . 5-speed
Final-drive ratio . . . . . . . . . . . . . . . . . . . . . . 3.89:1

| Gear | Ratio | Mph/1000 rpm | Max. test speed |
|---|---|---|---|
| I | 3.60 | 5.3 | 34 mph (6400 rpm) |
| II | 2.13 | 8.9 | 57 mph (6400 rpm) |
| III | 1.46 | 13.0 | 83 mph (6400 rpm) |
| IV | 1.07 | 17.7 | 113 mph (6400 rpm) |
| V | 0.73 | 25.9 | 122 mph (4700 rpm) |

### DIMENSIONS AND CAPACITIES
Wheelbase . . . . . . . . . . . . . . . . . . . . . . . . . . 94.5 in
Track, F/R . . . . . . . . . . . . . . . . . . . . . . . 58.2/57.1 in
Length . . . . . . . . . . . . . . . . . . . . . . . . . . . . 170.0 in
Width . . . . . . . . . . . . . . . . . . . . . . . . . . . . . 68.3 in
Height . . . . . . . . . . . . . . . . . . . . . . . . . . . . . 50.2 in

Ground clearance . . . . . . . . . . . . . . . . . . . . . . 4.9 in
Curb weight . . . . . . . . . . . . . . . . . . . . . . . . 2830 lbs
Weight distribution, F/R . . . . . . . . . . . . . . . 49.5/50.5%
Fuel capacity . . . . . . . . . . . . . . . . . . . . . . . 16.4 gal
Oil capacity . . . . . . . . . . . . . . . . . . . . . . . . . 5.3 qt
Water capacity . . . . . . . . . . . . . . . . . . . . . . . 9.0 qt

### CHASSIS/BODY
Type . . . . . . . . . . . . . . . . . . . . . . unit construction
Body material . . . . . . . . . . . . . . welded steel stampings

### INTERIOR
SAE volume, front seat . . . . . . . . . . . . . . . . . 50 cu ft
            rear seat . . . . . . . . . . . . . . . . . 12 cu ft
            trunk space . . . . . . . . . . . . . . . . 8 cu ft
Front seats . . . . . . . . . . . . . . . . . . . . . . . . . . bucket
Recliner type . . . . . . . . . . . . . . . . . infinitely adjustable
General comfort . . . . . . . . . . . . . . poor fair good **excellent**
Fore-and-aft support . . . . . . . . . . . . poor fair good **excellent**
Lateral support . . . . . . . . . . . . . . poor fair good **excellent**

### SUSPENSION
F: . . . . . . . ind, MacPherson strut, coil springs, anti-sway bar
R: . . . . . . . . . . . . . . . . . ind, semi-trailing arm, torsion bars

### STEERING
Type . . . . . . . . . . . . . . . . . . . . . . rack-and-pinion
Turns lock-to-lock . . . . . . . . . . . . . . . . . . . . . . . 3.8
Turning circle curb-to-curb . . . . . . . . . . . . . . . . 31.2 ft

### BRAKES
F: . . . . . . . . . . . . . . . . . . . 11.1 x 0.8-in vented disc
R: . . . . . . . . . . . . . . . . . . . 11.4 x 0.8-in vented disc
Power assist . . . . . . . . . . . . . . . . . . . . . . . . . vacuum

### WHEELS AND TIRES
Wheel size . . . . . . . . . . . . . . . . . . . . . . . 7.0 x 15 in
Wheel type . . . . . . . . . . . . . . . . . . . . . cast aluminum
Tire make and size . Pirelli Cinturato CN36, 215/60VR-15
Test inflation pressures, F/R . . . . . . . . . . . . . 31/34 psi

# Car and Driver Test Results

### ACCELERATION
| | Seconds |
|---|---|
| Zero to 30 mph | 2.4 |
| 40 mph | 3.7 |
| 50 mph | 5.4 |
| 60 mph | 7.5 |
| 70 mph | 10.1 |
| 80 mph | 13.0 |
| 90 mph | 16.9 |
| 100 mph | 21.7 |
| 110 mph | 28.3 |
| Top-gear passing time, 30–50 mph | 12.1 |
| 50–70 mph | 13.1 |
| Standing ¼-mile | 15.7 sec @ 87 mph |
| Top speed | 122 mph |

### BRAKING
70–0 mph @ impending lockup . . . . . . . . . . . . 190 ft
Modulation . . . . . . . . . . poor fair **good** excellent
Fade . . . . . . . . . . . . . . . . **none** moderate heavy
Front-rear balance . . . . . . . . . . . . poor fair **good**

### HANDLING
Roadholding, 282-ft-dia skidpad . . . . . . . . . . . 0.81 g
Understeer . . . . . . . . . . . . **minimal** moderate excessive

### COAST-DOWN MEASUREMENTS
Road horsepower @ 50 mph . . . . . . . . . . . . . 13.5 hp
Friction and tire losses @ 50 mph . . . . . . . . . . 7.0 hp
Aerodynamic drag @ 50 mph . . . . . . . . . . . . . 6.5 hp

### FUEL ECONOMY
EPA city driving . . . . . . . . . . . . . . . . . . . . **23 mpg**
EPA highway driving . . . . . . . . . . . . . . . . . . **36 mpg**
EPA combined driving . . . . . . . . . . . . . . . . . **27 mpg**
*C/D* observed fuel economy . . . . . . . . . . . . . **26 mpg**

### INTERIOR SOUND LEVEL
Idle . . . . . . . . . . . . . . . . . . . . . . . . . . . . . 56 dBA
Full-throttle acceleration . . . . . . . . . . . . . . . . 82 dBA
70-mph cruising . . . . . . . . . . . . . . . . . . . . . 73 dBA
70-mph coasting . . . . . . . . . . . . . . . . . . . . . 71 dBA

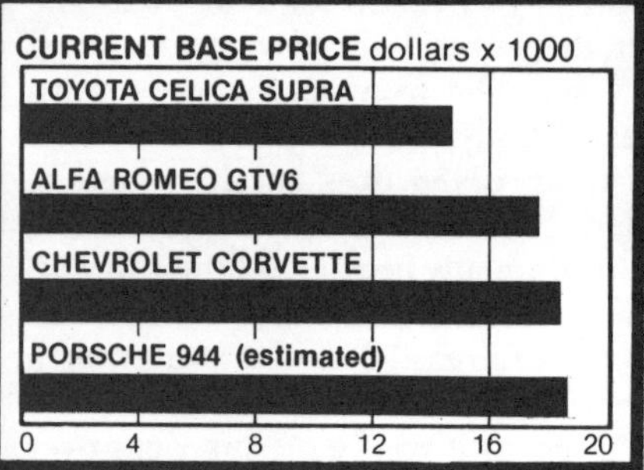

**CURRENT BASE PRICE** dollars x 1000
- TOYOTA CELICA SUPRA
- ALFA ROMEO GTV6
- CHEVROLET CORVETTE
- PORSCHE 944 (estimated)

(scale: 0 4 8 12 16 20)

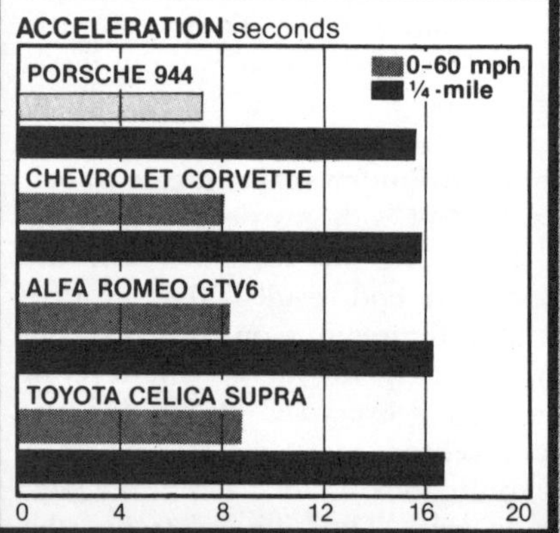

**ACCELERATION** seconds
Legend: ■ 0–60 mph ■ ¼-mile
- PORSCHE 944
- CHEVROLET CORVETTE
- ALFA ROMEO GTV6
- TOYOTA CELICA SUPRA

(scale: 0 4 8 12 16 20)

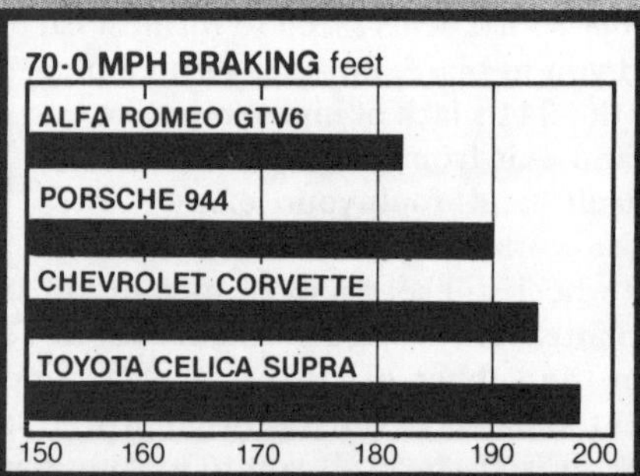

**70-0 MPH BRAKING** feet
- ALFA ROMEO GTV6
- PORSCHE 944
- CHEVROLET CORVETTE
- TOYOTA CELICA SUPRA

(scale: 150 160 170 180 190 200)

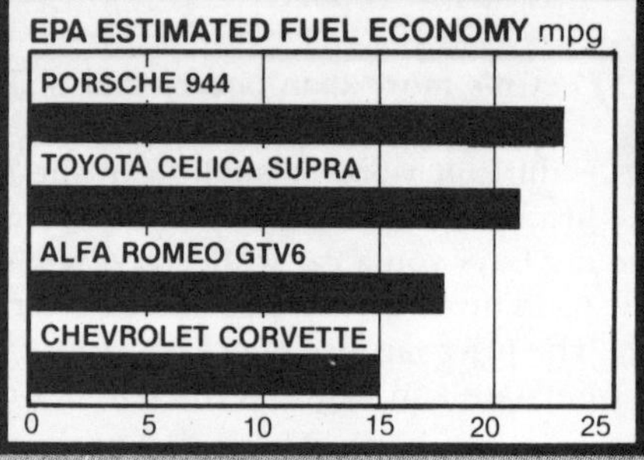

**EPA ESTIMATED FUEL ECONOMY** mpg
- PORSCHE 944
- TOYOTA CELICA SUPRA
- ALFA ROMEO GTV6
- CHEVROLET CORVETTE

(scale: 0 5 10 15 20 25)

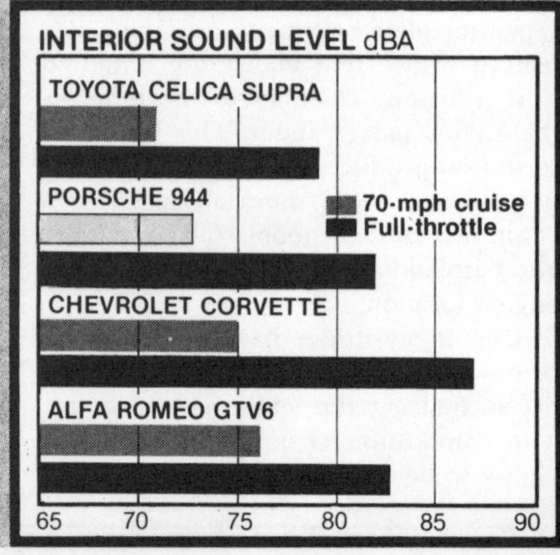

**INTERIOR SOUND LEVEL** dBA
Legend: ■ 70-mph cruise ■ Full-throttle
- TOYOTA CELICA SUPRA
- PORSCHE 944
- CHEVROLET CORVETTE
- ALFA ROMEO GTV6

(scale: 65 70 75 80 85 90)

# 24 Porsche 944

Engineered for maximum performance, the design of the Porsche 944 does not lend itself to ease and speed of assembly. We cannot increase our production schedule because to do so would mean to compromise the outstanding quality and performance that created the exceptional demand for this automobile in the first place.

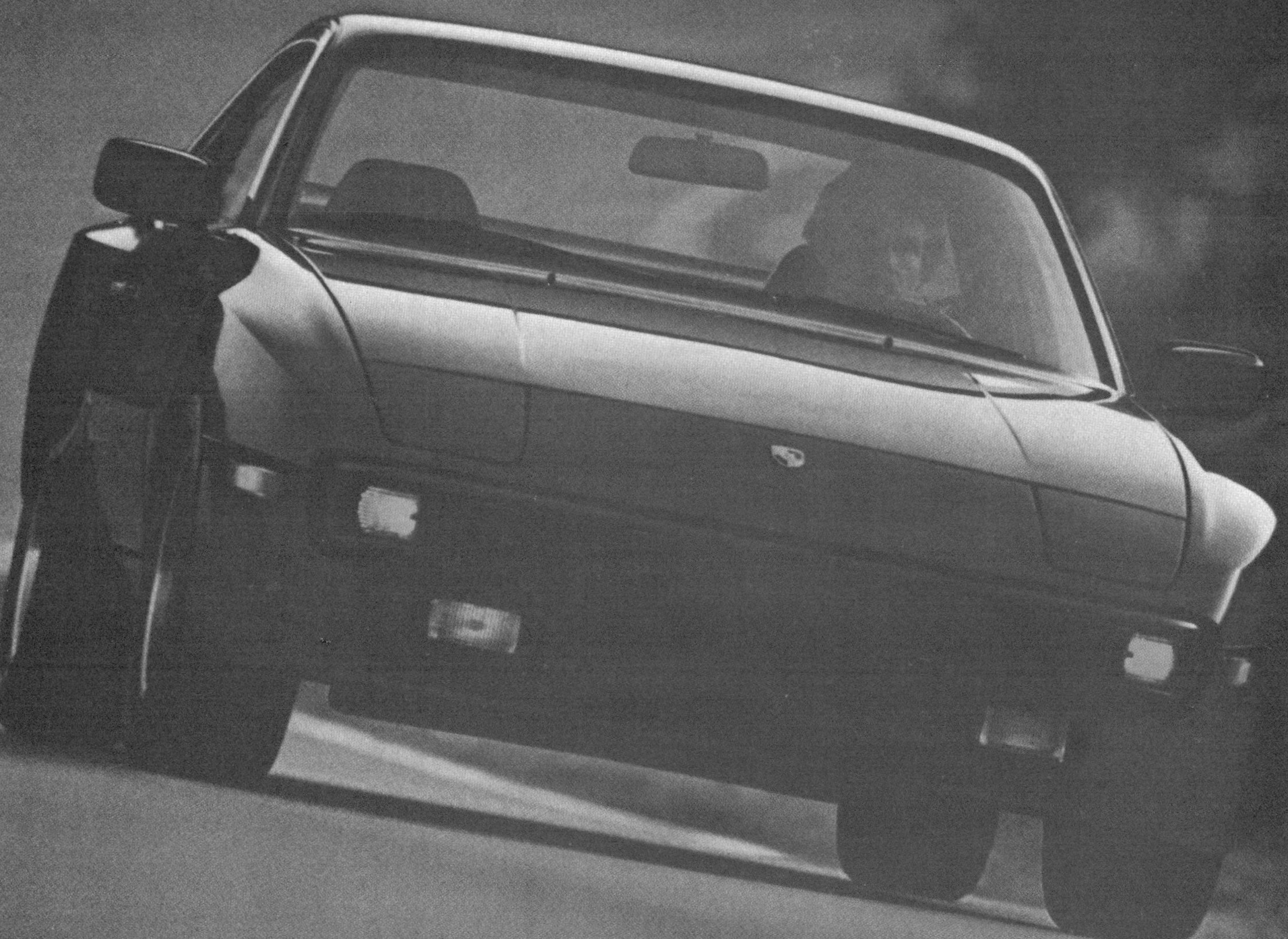

The Porsche 944 accelerates from 0 to 50 mph in 5.9 seconds. Top speed is 130 mph. Maximum torque is achieved as low as 3000 rpm.

Behind these impressive numbers is one of the largest, most powerful 4-cylinder engines in production: 2.5 liters, 143 hp.

To counter the vibrations an engine of this magnitude would normally produce, Porsche engineers have incorporated a system of counter-rotating silencer shafts and special engine mounts (actually mini shock absorbers).

The result: a 4-cylinder power plant that runs with the smoothness of a 6. Building the engine is, by necessity, a slow process, but the result is a high performance automobile.

Porsche's exclusive transaxle design provides an almost perfectly balanced front/rear weight distribution. This design requires drive train elements to be constructed as a single unit, instead of as individual components. Difficult to assemble, it contributes to the exceptional handling, cornering and directional control unique to Porsche.

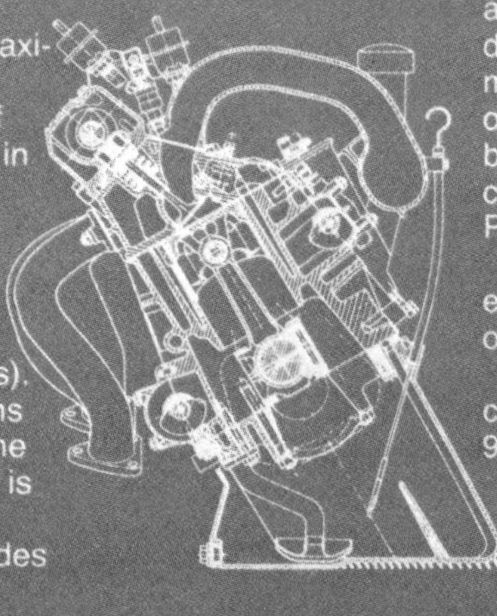

From its internal mechanics to its ergonomics, every aspect of the Porsche 944 is designed to optimize performance.

To insure high performace and quality, every car is inspected, every engine is tested, and every 944 is run on the open road prior to shipment.

The Porsche 944 is the antithesis of the mass produced automobile. But then, at Porsche, excellence is expected. **PORSCHE + AUDI**
NOTHING EVEN COMES CLOSE

# SPORT

EDITED BY LARRY GRIFFIN

# IMSA GT Confrontation: Ford Mustang versus Porsche 935

*A performance test of the two hottest GT racers in the old neighborhood.*

BY LARRY GRIFFIN

The tidy shadows of the Lola T600 and other GTP entries loom up on Group 5.

• Kings may die and be succeeded, and eras may pass and be replaced by others, but few of either will be mourned as much as the last hot breath of the Porsche 935.

We tested a 935K3 in the March 1980 issue, and Patrick Bedard, who actually bugged out blood vessels with the effort of driving the thing, came away feeling that the K3 might be an even more remarkable accomplishment than the Great Pyramid. The car he drove belonged to the Whittington brothers, and it had won Le Mans in June of 1979, just as Germany's Kremer brothers had intended when they built it. At the time, the 935 was the undisputed ruler of GT racing around the world, a domination that would continue until early 1981. Changes in the rules and growing interest from other manufacturers have changed everything, now.

Lola is building coupes for the new, full-race GTP category—which is intended to encourage flat-out competition in the manner of the old prototype days of Porsches, Ferraris, and Alfa Romeos, and new versions of mothballed dreams are on the way from a

PHOTOGRAPHY BY ROBERT HARMEYER, JR., AND AARON KILEY

half-dozen mightily muscled sources. Lola has already blown the gleaming patina right off Porsche's long-nurtured image of invincibility, and to make matters even more unpalatable to Stuttgart and Jo Hoppen, Porsche's competition czar in the United States, Ford has flagged down the IMSA express and hopped aboard with a European-built, Mustang-bodied, flyweight turbo flyer that's already snagged two outright wins in this year's Camel GT series.

It would be fair to say that Porsche is ''Hoppen'' mad. After years with the rules all its own way, the German juggernaut is so peeved that its parts bins

and support trucks have become suddenly much more difficult to find for teams that had been raking in glory since time immemorial.

Ford, on the other hand, is fresh back into racing and out for everything it can get. The quick way to fame was to reskin a suitable existing car and punch it into the winner's circle with as much fanfare as possible. Listen closely and you can hear the blaring brass all the way from Detroit. Put the music and the good spirits down to Walter Hayes and Michael Kranefuss, respectively the prime movers in Ford's reorganized PR department and new-in-1980 Special Vehicle Operations branch.

Despite its quick and encouraging success, the ''Mustang'' racer can't be much more than a stopgap against the same GTP cars that have so incinerated Porsche's peace of mind. Still, with the GTP cars in their early stages of development, both the Porsche and the Ford are certainly capable of winning, and you should not assume that either is anything but incredibly fast.

To support this view, we offer this test of both.

They are very different animals, each deadly in its way. The basic Porsche 935 Twin Turbo is a well-known quantity: air-cooled 3147cc engine in the back, somewhat antsy handling all the way around, terrific brakes no matter what, and about ten zillion horsepower to convert to forward motion. Anything with 720 to more than 800 horsepower and a rakish bulk of only 2430 pounds is enough to send any but the truest of the Mad GT Bombers back into the fetal position. John Paul, Sr., and John Paul, Jr., are still perambulating around upright, so it stands to reason that, since they drive this particular 935, they must qualify among the hierarchy of the Mad GT Bombers. The baby-blue JLP Racing 935 is one of the fastest ever built, but far from the most expensive at a thrifty, home-built $110,000 FOB the JLP shop (via the junkyard where the shell was found). The secret to this car's fearsomeness lies in its tubular construction, which gets away from Porsche's usual unit-body forays. The Pauls have built a very rigid 935 that carries its weight lower than your usual 935, thus allowing the bodywork to cling closer to the ground, where it nibbles around the fresh edges of ground effects for production-based racers. The result is a 935 that likes its cornering more than most.

The new, pint-sized bully on the block is perhaps even more adept at twinkling its toes. The Mustang is nearly 300 pounds lighter than the JLP 935, an advantage allowed by the rules because of its much-smaller-displacement 1745cc Cosworth BDA engine; but it, too, is massively turbocharged. The Miller-sponsored Mustang, run for Ford driver Klaus Ludwig by Bill Scott Racing, also has four valves for each of its four cylinders, whereas Porsche's six combustion chambers inhale and exhale through only two valves apiece. The Mustang's tubular aluminum chassis sits well down near the ground, the whole body sunk to about Salton Sea level in order to lower the center of gravity and cut wind resistance. Ford has also made a substantial attempt to take advantage of today's ground-effects technology by carefully smoothing the underside of the car, even to the point of enclosing the live rear axle, and by creating aerodynamic tunnels within the zany body. And they said only drag racing had funny cars!

The Mustang is a lower and wider car than the 935. Its lighter weight makes it more agile. Furthermore, the Mustang's front-mounted engine is positioned so far back under the windshield that it's effectively a mid-engined car with a much lower polar moment of inertia than the Porsche. This makes the Mustang capable of being easily squeezed into inadvertent holes in traffic. The Porsche, however, is restricted to more regimented handling behavior by virtue of its chassis layout. The Ford's smaller power output (depending on boost, from 530 to 620 horsepower) also makes it more drivable in quick-response situations.

As you can see, this confrontation is a classic in every respect, and we can only lament the fact that it didn't come to pass before the GT cars began to lose their foothold on the top rung.

Klaus Ludwig used to win virtually every German Group 5 Championship race in Kremer 935s before switching to the Ford persuasion in Europe, and now America, too, but the remarkable young John Paul, Jr., is the only driver to have sampled both recently. He did so at the joint request of Ford and *Car and Driver*. Ford wanted to see if he would indeed make a good running mate for Ludwig now, and if he could perhaps even replace the German in whatever effort Ford deems suitable for next year. *Car and Driver* wanted a full test on both cars and a chance to watch the much-discussed new talent in action.

We discovered something interesting in addition to adding two significant new sets of performance numbers to our burgeoning reference files. The Mustang and its fearsome-foursome little screamer of a motor, puffed up as it is by a single turbo while the Porsche has two, is capable of getting off the line *faster*. It actually enjoys a one- or two-tenths-of-a-second advantage to as much as 50 mph. But . . . from there up, the 935 begins to haul the mail: zero to sixty in 4.6 seconds versus 5.5 for the Mustang, zero to one hundred in 7.9 versus the Ford's 9.2, and the quarter-mile in 12.3 seconds at 136 mph while the Ford could turn ''only'' 12.9 at 127. The stopping capabilities were virtually

dead even, and the Mustang pulled out only a slight advantage on our skidpad.

As you look at the cars' specifications, notice the difference in rear suspensions (Ford's live axle versus Porsche's semi-trailing arms) and consider these John Paul, Jr., observations on the cars: "The Ford is a very nice car, very controllable. You have to be a little more careful with the Porsche. Just a little bit of oversteer seems to be the fast way with it, so what you're doing is steering with the rear wheels. Our tube-framed 935 is very stiff in the front, and this gets rid of the inherent understeer 935s have had when first turning in. It just digs right in. The turning-in capability of the Mustang seems to be very good, because it has so much of its weight on the front wheels. Exiting the turns, it's very easy to induce power oversteer, maybe because it's lighter in the rear. The only thing I didn't particularly care for was the brakes, but they seemed to hold up in the Norisring race. The pedal is just real mooshy all the time, like it's not 100 percent.

"The Mustang has hardly any lag for a turbocharged motor, which is very surprising. It comes off the corners well, although it doesn't seem to have as much thrust as the 935, but when you back out of the throttle and then get back on it, the boost comes back instantly. In the midrange it makes up for the loss of bottom end, which is very helpful at a lot of tracks."

From out of the mouth of a veritable babe, the definitive word on today's fastest production-based road racers. They may be dying, but damned if they're going to rest in peace. ●

## Vital Statistics

| | Miller Mustang | JLP Racing 935 |
|---|---|---|
| price | $250,000 | $110,000 |
| engine | turbocharged and intercooled 1745cc 4-in-line, water-cooled cast-iron block, aluminum head | turbocharged and intercooled 3147cc flat 6, air-cooled aluminum crankcase, barrels, and heads |
| valve gear | belt-driven double overhead cams, four valves per cylinder | chain-driven single overhead cam, two valves per cylinder |
| turbocharger(s) | AiResearch TB04S | 2 KKK K-30 |
| waste gates | 2 Porsche | 2 Porsche |
| intercoolers | 2 AiResearch, air-to-air | 2 AiResearch, air-to-air |
| maximum boost setting | 19.8 psi (race), 23.5 psi (qualifying) | 21.0 psi (race), 26.5 psi (qualifying) |
| compression ratio | 6.8:1 | 6.0:1 |
| horsepower @ rpm | 530 @ 9000 (race), 620 (qualifying) | 720 @ 7500 (race), 800+ (qualifying) |
| torque, lbs-ft @ rpm | 300 @ 6400 (race), 350 (qualifying) | 525 @ 5000 (race), 580+ (qualifying) |
| redline, rpm | 9000 | 8000 |
| transmission | 5-speed Getrag | 4-speed Porsche 930 |
| final-drive ratio | 4.38:1 | 5.13:1 |
| differential | 85% limited-slip ZF | none (titanium spool) |
| steering | rack-and-pinion, 1.3 turns lock-to-lock | rack-and-pinion, 2.0 turns lock-to-lock |
| front suspension | ind, MacPherson strut, coil springs, anti-sway bar | ind, MacPherson strut, coil springs, anti-sway bar |
| rear suspension | rigid axle, 4 trailing links, transverse Watt linkage, coil springs, anti-sway bar | ind, semi-trailing arm, coil springs, anti-sway bar |
| tires | Goodyear Eagle, F: 23.5 x 11.5-16; R: 27.5 x 13.5-19 | Goodyear Eagle, F: 23.5 x 10.5-16; R: 27.5 x 14.5-19 |
| wheels, in | BBS modular, F: 11.5 x 16; R: 15.0 x 19 | BBS modular, F: 11.0 x 16; R: 16.0 x 19 |
| brakes | 12.0 x 1.2-in vented disc | 13.0 x 1.9-in vented disc |
| curb weight, lbs | 2140 | 2430 |
| weight distribution, F/R, % | 50/50 | 44/56 |

## Car and Driver Test Results

| | Miller Mustang | JLP Racing 935 |
|---|---|---|
| acceleration, sec: | | |
| 0 to 30 mph | 2.4 | 2.6 |
| 0 to 40 mph | 3.2 | 3.4 |
| 0 to 50 mph | 3.9 | 4.0 |
| 0 to 60 mph | 5.5 | 4.6 |
| 0 to 70 mph | 6.5 | 5.3 |
| 0 to 80 mph | 7.2 | 6.4 |
| 0 to 90 mph | 8.0 | 7.1 |
| 0 to 100 mph | 9.2 | 7.9 |
| 0 to 110 mph | 10.7 | 9.4 |
| 0 to 120 mph | 12.0 | 10.5 |
| 0 to 130 mph | 13.3 | 11.7 |
| standing ¼-mile | 12.9 @ 127 mph | 12.3 @ 136 mph |
| top speed, mph | 180 (Atlanta gearing) | 225 (Le Mans gearing) |
| braking, 100–0 mph, ft | 360 | 355 |
| 70–0 mph, ft | 177 | 174 |
| roadholding, g | 1.26 | 1.20 |
| typical racing fuel economy, mpg | 3–4 | 3 |

# Red Speed

*Gauging the good, better, best of the all-American road machines
against the ultimate standard.*

• Five men, good and true, set out on what seemed like a mission from God: to live with three of the fastest (and reddest) American GTs for two weeks, then sit down to haggle over the findings. A Porsche 928 was anointed to serve as the group's spiritual inspiration and silver-plated yardstick. Our data-processing machine would systematically sort the raw test results; every annotation of a squeak, rattle, or death-defying maneuver would be culled from the drivers' logs and fed into the judgment hopper; finally, each editor, armed with half a month's worth of fact and fancy, would sequester himself to fill out his own Grand Tally Sheet of Subjectivity. The good-better-best final verdicts would spring forth from the voting much as the Ten Commandments fell out of the sky into Moses' lap. The all-American-GT champion would emerge, and peace would transcend all evil.

At least that was the script. Reality, however, has a habit of going its own dastardly way. We did have a ball hammering these four thunderbuggies up California's Mount Palomar

and down the Ortega Highway. Along the way, we found a lot to love about the Camaro, Mustang, and Trans Am. But we also found a surprising number of traits even a mother would abhor. Toward the end, the Trans Am eked out a one-point victory over the Mustang (plus five points over the Camaro) in the subjective ratings, but that same Pontiac marched smartly to a *last*-place finish in our fun-to-drive ranking. What's more, none of the three contenders could drum up a majority when each of the five editors was pressed for a simple "Best GT" nomination.

Sad to say, we have for you no clear winner. Our findings tell us the silver-bullet solution to your GT needs is yet to be made in Detroit. This is not to suggest that the test was a bust; before we sign off, we'll be happy to confide our most intimate experiences with the four racy-red V-8 GTs. Who knows? This comparo test might tip your balance toward the two-plus-two of your dreams. Then again, it could be the last straw before you rush out and order a four-cylinder station wagon.

**Chevrolet Camaro Z28:** We or-

dered our Camaro with the fastest powertrain available—5.0-liter TBI V-8, Turbo Hydra-matic transmission, 3.23:1 axle ratio—even though most of us at *C/D* prefer a GT car's shift lever to be more than a hand rest. To examine the other side of the coin, we ordered the Firebird with the hairiest manual-transmission powertrain offered: the 5.0-liter V-8 (inhaling through a four-barrel carburetor, exhaling through a more restrictive catalyst), the venerable Borg-Warner T-10 four-speed, and the same 3.23:1 "performance" axle ratio. Truth to tell, neither powertrain stirred anyone's soul to ecstasy.

We could carp all day long about catalytic converters and the enemy in Washington, but the essential problem is underhood impotence stacked against 3400 pounds of dead weight. Five liters is an ample piston displacement in the era of $1.50-per-gallon energy. And most manufacturers have found benefits from fuel injection other than simply draining the gas tank expediently (with a capacity of sixteen gallons and a test-trip efficiency of 12 mpg, both the Camaro and the Firebird needed pit stops every 192 miles). Chevy is currently building five-speed manuals into Chevettes and four-speed automatics into Impalas, so why should the fresh-tech Camaro limp through life with ineffective and unfun powertrains?

When you drive a Z28, there is one engineering breakthrough that slaps you right in the face: this Camaro is *not* a committee car. The shock valving is so tight that you feel pebbles on the pavement as you back out of a parking space. And the steering response is uncannily importlike to the touch. When your wrists move the wheel, the Z28 responds with the keen linearity that used to be the exclusive province of the better Porsches, Ferraris, and Lotuses of the world. This is not just a lack of the usual on-center null (dead zone) that Detroit has championed for years, but a newfound straight-shot steering response that starts in your garage and runs into the 0.8-g realm. You turn, it turns. There is no slack.

The serpentine switchbacks up and down Mount Palomar demonstrated a clear difference between Camaro and Firebird handling: the Camaro cuts and thrusts with the Porsche, while the Trans Am is noticeably slower on the uptake. To find out exactly why, you have to penetrate the GM hierarchy. Once inside the Chevrolet Division's perimeter and at the crux of the matter, you'll find Fred Schaafsma, a chassis-development engineer who bucked his personal preferences all the way through the convoluted system into production. When you tell Schaafsma the Z28's ride over expansion joints makes your teeth ache, he replies that there are base Camaros and a Berlinetta upgrade for the comfy-cruising set, while the Z28 was planned from the start as the bare-knuckle street fighter. One natural foe was the Trans Am, so Schaafsma spent the money and effort to stiffen the Z28's step until he was sure it would walk around the Pontiac, and just about any import you could name, in handling. As it stands today, the Z28 and T/A share plenty—powertrains, basic body structures, wheel width, tires, even anti-sway-bar diameters—and Schaafsma will speak of the key chassis differences only in roundabout terms: stiffer rear springs, an unspecified subtlety in the rear suspension, and six particular braces and bushings in the front end. These latter pieces make the Z28 feel steel-bowstring taut over bumps, while the Trans Am is much more resilient. They also help make it the demon of the slalom run, where the Trans Am is merely excellent. Both cars still have a way to go in steering feel, however. The Porsche along for the ride had a higher effort at the wheel rim and a far more tactile touch with the road, two traits we admire in any context.

Some other aspects of the driver's interface with the chassis also need attention. The extra-cost Conteur driver's seat is as comfortable as a pincushion: the too-aggressive under-thigh bolster acts like a leg tourniquet even in its most retracted position, and the steel wires that serve as the side bolsters' skele-

ton poke through painfully as cornering loads build. We prefer the Camaro's instrument panel to the Firebird's, but can't understand the nickel-sized gauge ports or the hard console edge that hates the driver's right thigh.

It should be obvious by now that we have an amalgam of emotions for the Z28. Our fun-to-drive favorite (even with the lackadaisical automatic transmission), it fell to second place in acceleration and all the way to the bottom in the subjective totals. The makings of a great car are here, but the development program is far from over.

**Ford Mustang GT:** The flier from Ford was crude and comparatively unwieldy in several of the decathlon events that composed this GT olympics, but in terms of sheer visceral appeal, it's right up there with the Porsche. Press on the Mustang's gas pedal, and great things happen. An authoritative growl from under the hood is accompanied by screeches of rubber at the back of the car. This Mustang is at the moment the quickest machine made in America, and our internal sources at the Ford Motor Company suggest that efforts are afoot to keep Mustangs and Capris that way.

The Mustang is hardly a one-dimensional quarter-mile specialist, but it clearly lacks the chassis to compete with either Porsche's finest or the latest science from GM. What the Mustang does enjoy deep within its four-year-old, Fairmont-derived underpinnings is an ideal basic size. It's ten inches shorter, three inches narrower, and a hefty 400 pounds lighter than the new GM kids on the block; as a result, it feels nimble and eager to get the job done. The body's comparatively blunt shape doesn't benefit from the recent intense aerodynamic-drag efforts, but there is a compensating benefit: the less-sleek Mustang offers the only real back seat in the test and the roomiest luggage compartment.

The Mustang's front seating is also a credit to the GT class. Instead of attempting to reinvent the sport seat in-house, as

Chevrolet (actually Fisher Body) has done with the Camaro, Ford shopped at the acknowledged orthopedic experts'—Recaro—and fulfilled its obligation once and for all.

The not-so-satisfying phases of Mustang life have to do with handling. The rack-and-pinion steering is one of those pinky-finger power jobs where actual road feel is not allowed. Down at the business end of the steering linkage, Ford has gone a bit skimpy on tire size, so the smallish Michelin TRXs have all they can do to generate a feeble 0.75 gs' worth of grip on the skidpad. The engineers have done a fair job tuning the Mustang's ride—judged better than the Camaro's, not as plush as the Trans Am's—and their roll-stiffness distribution has set the handling commendably close to neutral steer. One small problem that's left is that there is not enough anti-sway bar to flatten out the proceedings through life's ess-bends.

A bigger concern is a woefully inadequate rear suspension. What we have here is the Peter Principle applied to four-bar linkages. The Mustang's rear suspension started life in the Fairmont, where it worked just fine. But with 240 pounds-feet of torque multiplied by a manual transmission, those four stamped-steel links are burdened well beyond their level of competence. A set of windup limiters has been Band-Aided into the system for 1982, but they're only half as effective as a full fix would be. In left-hand sweepers, the gas pedal acts as a power-oversteer switch: put your foot down (in the lower gears) and the tail slips out. That smooth two-step unfortunately turns into a jitterbug in right-hand bends, where power hop conspires with the transmission to make life difficult. The Mustang's box shifts better than the Pontiac's and the Porsche's, but on the road it acts convincingly like a five-speed with a missing third gear. In mid-speed right-handers you're confronted with a nasty dilemma: to power-hop your way around in second, or to chug through in third?

We do expect some attention to these and other problems

in the Mustang as Ford tools up the guts to go behind the GT badge, but in the meantime, there's no denying the terrific value here. You can still roll a hot Mustang (or Capri) out the door for less than ten grand. In terms of power-to-weight ratio for a price, there's nothing that can touch the Mustang from GM, Germany, or even Japan.

**Pontiac Firebird Trans Am:** This test makes one thing perfectly clear: Trans Ams are not what they used to be. The proud bird on the hood now broods over Chevrolet-built engines. The mean and macho WS6/7 has been clipped back with refinement, its ride/handling balance skewed noticeably more toward comfort than the rival Chevrolet. And while the

# Chevrolet Pacing, As You Like It

• Simplicity is a virtue of great complexity, as anybody who has built cars will tell you, but it's often difficult to convince the guys in the leather chairs that the simple approach is the best way to make a big splash. They usually seem to feel that the more flash, the bigger the splash. Recent Indianapolis 500 pace cars have been so highly modified that they were also highly likely to provide one big flash and altogether the wrong kind of splash.

Providing pace cars for the Indy 500 is a risky business. This is, after all, the most-watched automotive event in the world—a third of a million people mashed fanny to forepart in Indiana and tens of millions more watching and listening around the world. So here come you, the Michigan mogul, already under fire for falling sales and so-so quality control, and you're represented, as anybody still alive on Memorial Day weekend can plainly see, by this pace car of yours, and it's got to start and run and go flitting in and out of the pits on command to lead all these race cars around in front of all these people. You don't want it to swallow its starter or hiss out its coolant or bolt for the wall or sprinkle its engine all over the front straight. Extremely bad form, you know.

If you happen to be Chevrolet and you've bid yourself into providing a matched set of Camaro Z28s to pace the 1982 Indy 500, you take one look at the overengineered recent efforts of others and you get John Pierce on the phone. John Pierce is in Product Promotion Engineering at your engineering center, and you say, "John, build us a pace car, not a prototype hand grenade." He dives headlong into your corporate book of performance parts, appropriately entitled *Chevrolet Power*, and comes out with a pair of Z28s boasting 50 percent more horsepower than stock, but no bad habits.

What Pierce has put together is Chevy's proven smallblock 5.7-liter racing V-8 engine, an animal as reliable as sun in the Sahara and further tamed with your basic streetoption crossfire fuel injection, which proves surprisingly compatible with the hefty and yet feathery mechanicals underneath it. The familiar all-aluminum racing block is fitted with aluminum heads, big valves, hydraulic lifters, the high-output cam first developed for the 327s almost twenty years ago, forged aluminum pistons, and heavy-duty crank and connecting rods. The compression ratio is 11.0:1 (versus the stock 5.0-liter's 9.5:1) and output is up from 165 horsepower to an easy 250, estimated.

The catalytic converter has been left on the shop floor and the heavy cast-iron exhaust manifolds have been deleted in favor of thin-wall headers. The engine itself weighs 120 pounds less than its cast-iron counterpart, and the total saving means the pacing Z28's weight distribution comes in at very nearly 50/50. Optional four-wheel discs are used, and the stock Z28 suspension is retained. The single change in running attire is a switch from 215/65R-15 tires to shaved Goodyear P245/60R-15s. European-export headlamp covers and a 140-mph speedo are the only other additions of note beyond the roof lights, radio system, and an ABC TV camera hidden behind a tail lamp.

Cruising at 90 on the banked, glassy surface of the big speedway is effortless. The engine loafs, but it's snugly ready to rush the three-speed automatic forward at a touch. Danny Ongais took one of the cars out and declared a 120-mph average to be no more than a two-finger steering operation (even if his fingers are worth about a dozen of anybody else's). The shame of it is that this effortless performance is available to no one but pace-car drivers (at least not from the showroom). And these all-aluminum engines look *so* pretty and tidy and clean. For once, simplicity hasn't required great complexity. This obviously will have been a great comfort on race day for those of you who occupy Chevrolet's leather chairs. —*Larry Griffin*

old T/A was a flapped and spoilered war bird, the new species is more the pretty peacock. Its long, shapely snout and sleeker roofline slide through the air without a ruffled feather.

As with the Chevy in this test, the Firebird has problems. Something is desperately wrong when a Nissan Stanza econosedan beats the ballsiest four-speed T/A to 60 mph.

And almost all the basic traits that have to do with using the Firebird's stick shift are miserable. A courtesy light is quite inconveniently located in the already tight space over your clutch foot; nine shifts out of ten, you snag the sole of your shoe. Heel-and-toeing is difficult, if not impossible. The transmission linkage is a high-effort, low-precision device

# Aerodynamic Annotations

• Aerodynamics is all the rage in Detroit these days. The manufacturers have finally discovered the benefits of a clean shape, and they're trumpeting their achievements with the zealous fervor of new converts. Pontiac is even claiming that the new Firebird Trans Am is the world's slipperiest production car, with a drag coefficient of 0.31. While we can't name a car with a lower Cd, we are a bit skeptical about this figure. We have a basic distrust of drag coefficients because they inevitably mislead. Furthermore, GM has never reconciled the Trans Am's 0.31 drag coefficient with the Camaro Z28's 0.37 figure.

That's a large difference for two very similar cars. The Trans Am does have retracting headlights, which are not only sleeker than the Camaro's but, more important, also allow a lower hood line, by nearly two inches at certain points. The Trans Am's flush wheels are also an advantage, albeit a small one (the Cd benefit is less than 0.01, according to sources within GM). In other respects, however, there's not much to choose between the two cars. The front-spoiler designs are completely different—the Z28's is on the leading edge of the bumper, while the Trans Am's is just forward of the front wheels—but neither arrangement is clearly superior. Ditto the rear spoilers, which, though contoured differently, are virtually identical in effectiveness (both trim Cd by less than 0.01). It's hard to see where the total of these differences adds up to a 16 percent advantage for the Trans Am.

Differences in aerodynamic-drag coefficients are often a result of wind-tunnel inconsistencies, but both of these cars were tested at the same GM facility in Warren, Michi-

gan. This doesn't eliminate all the wind-tunnel variables, however. Most cars have less drag when they sit lower. Recognizing this, Pontiac says that the 0.31 figure was obtained at "EPA coast-down height," though it makes no attempt to explain what that means. Also, production tolerances in springs and other suspension components may be used to a manufacturer's advantage during testing.

Another significant source of aerodynamic variation is body-panel fit. Close-fitting, flush seams are better than wide, protuberant ones. Sloppy fits not only upset airflow over the car, but can allow harmful internal airflow as well. For example, eliminating airflow through the duct in the Z28's front spoiler improves its Cd by 0.02. In fit as in ride height, two cars can be within production tolerances, yet be quite different to the wind.

Our coast-down testing is also affected by these variables, but at least we test production cars, not hand-massaged prototypes. Furthermore, the coast-down test is a real-world situation, with a moving car, a fixed ground plane, and rolling tires; in a wind tunnel, everything but the airstream is stationary. In testing the Trans Am and the Z28 back-to-back, we found that the Pontiac required 8.0 horsepower to overcome aerodynamic drag at 50 mph, while the Z28's aero horsepower is 8.5. Since the frontal areas are nearly identical, this difference has to be a result of the Trans Am's lower drag coefficient.

This doesn't mean that the 0.31 drag-coefficient claim is less suspect, but it does prove—to our satisfaction, at least—that the Trans Am is a significantly cleaner shape than the Z28.  —*Csaba Csere*

that's annoying to use. With no fifth in the box, on the highway you feel you're stuck in an intermediate gear.

The four-barrel V-8 is a torquey and responsive prime mover at the low end, but a run through the gears is really just four short spurts instead of one sustained burst of enthusiasm. Thanks to the restrictive pellet-type catalyst packed into its pipes, the LG4 engine's power curve wilts on cue right after its 4000-rpm peak. You can force matters to the 5000-rpm redline—in fact, you can cruise at that speed in top gear (116 mph) on about two-thirds throttle, thanks to the tidy aerodynamics—but there's so much driveline coarseness fed up and into the car at higher rpm that it rarely seems worth all the commotion. Will the Japanese have to bring out six-speeds to convince GM that a nice, easy-shifting, properly ratioed five-speed is part of doing business these days?

There *are* reasons to celebrate the new Firebird. The exterior sculpturing is an absolute knockout, and we hope and pray the sales department doesn't ruin the artwork with its usual demands for scoops, stripes, and more bird droppings. The chassis is a satisfying accomplishment, admittedly not as crisp

## Vital Statistics

| | price, base/as tested | engine | power/torque | transmission/ axle ratio | curb weight, lbs |
|---|---|---|---|---|---|
| **CHEVROLET CAMARO Z28** | $9700/$13,191 | V-8, 305 cu in (5001cc), iron block and heads, 2x1-bbl throttle-body fuel injection | 165 bhp @ 4200 rpm/ 240 lbs-ft @ 2400 rpm | 3-speed automatic with lockup torque converter/3.23:1 | 3440 |
| **FORD MUSTANG GT** | $8308/$10,167 | V-8, 302 cu in (4942cc), iron block and heads, 1x2-bbl carburetor | 157 bhp @ 4200 rpm/ 240 lbs-ft @ 2400 rpm | 4-speed/3.08:1 | 3000 |
| **PONTIAC FIREBIRD TRANS AM** | $9658/$12,836 | V-8, 305 cu in (5001cc), iron block and heads, 1x4-bbl carburetor | 145 bhp @ 4000 rpm/ 240 lbs-ft @ 2000 rpm | 4-speed/3.23:1 | 3400 |
| **PORSCHE 928** | $39,500/$43,840 | V-8, 273 cu in (4474cc), aluminum block and heads, Bosch L-Jetronic fuel injection | 220 bhp @ 5250 rpm/ 265 lbs-ft @ 4000 rpm | 5-speed/2.75:1 | 3340 |

## Test Results

| | acceleration, sec | | | | | top speed, mph | braking, 70-0 mph, ft | roadholding, 282-ft skidpad, g |
|---|---|---|---|---|---|---|---|---|
| | 0-60 mph | 0-100 mph | 1/4-mile | top gear, 30-50 mph | top gear, 50-70 mph | | | |
| **CHEVROLET CAMARO Z28** | 8.6 | 27.3 | 16.4 @ 83 mph | 3.8 | 5.9 | 116 | 201 | 0.81 |
| **FORD MUSTANG GT** | 8.1 | 24.3 | 16.2 @ 86 mph | 9.5 | 10.5 | 125 | 213 | 0.75 |
| **PONTIAC FIREBIRD TRANS AM** | 10.6 | 31.4 | 17.5 @ 80 mph | 6.1 | 6.7 | 116 | 186 | 0.81 |
| **PORSCHE 928** | 6.8 | 19.0 | 15.1 @ 91 mph | 7.5 | 7.9 | 135 | 180 | 0.81 |

as the Camaro's, but a maneuverable piece just the same.

Unfortunately, something went haywire in the interior. The base front seats are essentially J-car components—not bad for a compact, but hardly up to snuff in a $13,000 GT. Pontiac does, make that did, offer a special Recaro package that unfortunately tied the T-top roof into the bill of materials, running the tab for this one option to $2486. The 2000 that were planned for production through May were promptly snapped up anyway, so we're back to the J-car seats, which aren't rated for the 0.81-g cornering capability of the new Trans Am.

Then there's the instrument panel. The far-out aerospace look was supposedly the goal, but in execution the Trans Am never quite breaks out of the Piper Cub realm. Allen-head screws are sprinkled about unconvincingly, and the instrumentation is belittled by huge plastic bezels. The steering wheel, a nice, classic three-spoke design, suffers from the heavy-handedness of the plastic-allen-screw proponents. The view over the rim and down the steep hood is a joy to behold, however, particularly if you select a Firebird without the asymmetric "power" bulge.

| weight distribution, F/R | dimensions, in | | | | fuel tank, gal | suspension | | brakes, F/R | tires |
|---|---|---|---|---|---|---|---|---|---|
| | wheelbase | length | width | height | | front | rear | | |
| 55.8/44.2 | 101.0 | 187.8 | 72.1 | 49.8 | 16.0 | ind, MacPherson strut, coil springs, anti-sway bar | rigid axle, 2 trailing links, Panhard rod, torque arm, coil springs, anti-sway bar | vented disc/ vented disc | Goodyear Eagle GT, P215/65R-15 |
| 57.3/42.7 | 100.4 | 179.1 | 69.1 | 51.4 | 15.4 | ind, MacPherson strut, coil springs, anti-sway bar | rigid axle, 4 trailing links with windup limiters, coil springs, anti-sway bar | vented disc/ drum | Michelin TRX, 190/65HR-390 |
| 57.6/42.4 | 101.0 | 190.3 | 72.4 | 49.8 | 16.0 | ind, MacPherson strut, coil springs, anti-sway bar | rigid axle, 2 trailing links, Panhard rod, torque arm, coil springs, anti-sway bar | vented disc/ drum | Goodyear Eagle GT, P215/65R-15 |
| 50.3/49.7 | 98.4 | 175.1 | 72.3 | 51.7 | 22.7 | ind, unequal-length control arms, coil springs, anti-sway bar | ind, unequal-length control arms, coil springs, anti-sway bar | vented disc/ vented disc | Pirelli P7, 225/50VR-16 |

| maneuverability, 1000-ft slalom, mph | road horsepower @ 50 mph | | | interior sound level, dBA | | | fuel economy, mpg | | |
|---|---|---|---|---|---|---|---|---|---|
| | total | aerodynamic | friction | idle | 70-mph cruising | full throttle | EPA city | EPA highway | C/D 400-mile trip |
| 61.2 | 14.5 | 8.5 | 6.0 | 59 | 76 | 78 | 16 | 24 | 12 |
| 56.3 | 15.5 | 9.0 | 6.5 | 57 | 75 | 82 | 17 | 28 | 13 |
| 59.6 | 14.0 | 8.0 | 6.0 | 56 | 74 | 78 | 16 | 24 | 12 |
| 60.4 | 15.5 | 8.5 | 7.0 | 55 | 73 | 78 | 16 | 25 | 13 |

| Editors' Choice* | | power-train | brakes | byway handling | highway handling | ride | ergo-nomics | overall comfort | value | overall rating | fun-to-drive |
|---|---|---|---|---|---|---|---|---|---|---|---|
| | CHEVROLET CAMARO Z28 | 12 | 15 | 21 | 19 | 10 | 18 | 14 | 18 | 127 | 11 |
| | FORD MUSTANG GT | 19 | 13 | 14 | 16 | 14 | 15 | 18 | 22 | 131 | 10 |
| | PONTIAC FIREBIRD TRANS AM | 11 | 16 | 20 | 20 | 18 | 18 | 12 | 17 | 132 | 9 |
| | PORSCHE 928 | 23 | 25 | 25 | 23 | 19 | 24 | 22 | 15 | 176 | 20 |

*Five editors each rated all four cars subjectively, assigning a maximum of 5 points to each car in eight categories. The overall ratings are simply sum totals of each car's individual ratings. In the fun-to-drive category, five editors each *ranked* the cars from best (4 points) to worst (1 point).

In the American tradition, there are plenty of ways to spec out a Firebird; as a matter of fact, the powertrain from the Camaro Z28 in this test is also available at your local Pontiac dealer. If you find even that eminently resistible, there is a ray of hope for the future. We understand better seats, a five-speed transmission, and stickier tires will enter the Trans Am's options list within the year.

**Porsche 928:** The Porsche was a terrific reference in this endeavor because, in our collective opinion, it is nearer perfection than any other GT ever built. This is not to say that we're scolding Detroit for denying car freaks the $40,000 made-in-America exotica they so richly deserve. To the contrary, Camaros, Firebirds, and Mustangs are already expensive enough, but they clearly have room for improvement. We see the 928 as the right kind of goal for future efforts.

Where Porsche has them all licked is under the hood. Yes, it's powered by a very expensive, all-aluminum, overhead-cam V-8, but there's far more spirit here than simple money can buy. That engines are for making power is a fact of life at Porsche, but one that seems to be lost on most engineers in Detroit. When you step on the gas pedal, the 928 rushes smartly for the 6000-rpm redline with no timeouts to remind you that this is a gas-mileage motor, or a prime mover tuned only for the EPA dyno and the first 30 feet away from a stoplight.

Five nicely spaced transmission ratios help keep the underhood spirit alive, but here you find the 928's Achilles' heel. The 928's shift linkage generated widespread bad reviews to the extent that most of us are avowed 928-automatic fans.

Although the domestics are knocking on Porsche's door in handling, the 928 has a very distinct edge you won't find in the test results. It *feels* much better through the steering, because forces from the tire patches are delivered to your finger tips to tell you just what's going on. The 928's attitude approaching the limit is also far more sensitive to the driver's commands than the other three cars in this test. No matter how hairy the maneuver, there always seems to be a little agility left in reserve. Adjustments to the throttle generate an amazingly adroit response: lifting ever so slightly causes the nose to tighten its cornering arc a notch at a time without the usual slewing and sliding at the back of the car.

If we were chief engineers for a day at any domestic division interested in the GT business, the very first $40,000 out of the budget would go for a 928. What this car has is too special to be the private reserve of the ultrarich forever.

**Conclusion:** The perfect GT is yet to be built. The Porsche 928 has just about everything necessary to qualify except an affordable price. We could be quite happy with the Mustang's seating and acceleration, the Camaro's handling, the Trans Am's aerodynamics and ride comfort, if all these qualities could be neatly wrapped in one package. Such a car is clearly within the realm of possibility. And if it doesn't come from Detroit soon, you can be sure one will head our way from across the water. Just watch for an ominous speck growing in the glare from the Land of the Rising Sun.   —*Don Sherman*

The *C/D* testers: Shades of Mt. Rushmore.

scienced-out suspension helps make the 924T extremely forgiving. The Greens, both of whom are certified Solo I leadfoots, say the 924 responds much quicker than their old 914 racer, yet the consequences are progressive and predictable instead of violent. Only the steering feels wonky to me: the low spokes of the Porsche Design steering wheel accentuate the low position of the wheel itself, so I found myself driving with my elbows digging into my ribs. And the steering feels slow as well.

For all its sure-footedness, though, the Turbo doesn't feel very powerful. The engine has been updated to 1981 specifications, which means more sensors to determine ignition advance and a quicker build up of boost pressure. The dyno chart indicates more than 200 horsepower. But despite the soul-satisfying whistle and pop of a racing Porsche turbomotor, there isn't much thrust. A quick lap is a matter of rowing the sloppy five-speed like a maniac (necessary because of the shorter-ratio third, fourth, and fifth

# Automotion Porsche 924 Turbo

## *Looking good with the ultimate Porsche café racer.*

• This is a café racer. You know what a café racer is. It's a car that's speedy, meant to be raced in short bursts to and from profiling sessions at your favorite café. Of course, some people claim that café racers got their name because they're so uncomfortable that you're always racing to the next café to make a comfort stop.

These days, when a café racer isn't at the local bistro it's at a Solo I competition, a harmless weekend recreation in which a hundred or so people attempt to destroy their street cars on a road-racing circuit. After four 45-minute practice sessions spread over two days, Solo I competitors set a quick lap in one three-lap banzai charge.

This Porsche 924 Turbo café racer was built by Automotion (3535 Kifer Road, Santa Clara, California 95051; 408–736–9020), perhaps the foremost purveyor of aftermarket Porsche parts and certainly the outfit with the classiest catalog. Marj and Tom Green of Automotion worked with Carlsen Porsche+Audi in Palo Alto, California, to create this car as a rolling showcase for Solo I hardware. It incorporates every last trick in Automotion's three-dollar catalog, and some on top of that.

The Greens began this project by stripping this leftover 1980 924T to the last nut and bolt. Then the tide of aftermarket parts rolled in.

First the safety hardware: belts from Deist and a roll bar from Autopower. Then a Porsche Design steering wheel from Momo and a Recaro racing seat. For the exterior, swoopy front fenders and rear quarter-panels and an air dam from JAM Fiberglass, plus sixteen-inch BBS wheels and Goodyear tires

(slicks for the track, 50-series NCTs for the street). For the engine, complete blueprinting, a lighter flywheel, and Automotion's other competition accessories, including its Turbo Oiler, which primes the turbo's bearings with oil prior to start-up.

Maximum effort went into the suspension. Race-car builder Jon Milledge redesigned the suspension geometry and wheel rates. The location of the MacPherson struts was subtly altered to eliminate bump-steer and add anti-dive. Spherical rod ends replaced rubber bushings in the front suspension. H&H Specialties built a 25mm front anti-sway bar. Then stiffer springs and Bilstein shocks were fitted. The rear also got an H&H anti-sway bar and Bilsteins, while Sway-A-Way built new 26mm torsion bars and ride-height adjustment plates. Milledge then fabricated a 935-style camber box to locate the rear suspension's inner pickup points so camber could be changed without affecting toe-in. In addition, the dozen or so bushings that help locate the rear suspension (increased by Porsche from four in 1980 to fight vibration) were replaced by bushings of Automotion's design.

Naturally, no sane person would ever do this to his car. As Milledge says, the suspension is better than those of most flat-out, full-bore IMSA race cars. But then Solo I enthusiasts, like most racers, tend to have more money than judgment.

The suspension does work, though. Once I'd slapped on my Bell and pointed the Turbo's nose toward Turn One at Sears Point Raceway, I might as well have had a race car under me. Equal weight distribution, both fore-and-aft and side-to-side, as well as a

gears installed by Automotion for competition) and carrying as much speed as possible through the corners.

But you want to know what this car is like to drive on the street. Well, no sane person would do so. Tiny twin shock absorbers from Porsche's racing catalog can't control the engine's vibes, while the rear suspension always sounds as though it's been invaded by a miniature hailstorm. And with so much vibration, there's always the chance that something will break or fall off.

Yet once you abandon your sanity and slide a little foam underneath you, you can drive this car on the street and smile about it afterward. You can jack up the ride height to clear the road's high spots, and there's plenty of suspension travel to cope with moderate bumps. Sure, this is still a race car on the street and it wears you out like a motorcycle, but it is definitely drivable.

And I figure that most of the charm of a café racer is the fact that it really *is* a race car on the street. It represents pure car dynamics, with none of that rational stuff like comfort or reliability. And the way it looks, thanks in no small part to the voluptuous fiberglass, drives me wild.

So let's be honest. The whole concept of a race car on the street is too outrageous for rational analysis. Who cares how this car drives? The best thing about a café racer is the way it looks from the veranda of your favorite bistro.

*—Michael Jordan*

Porsche 917 Turbo

Porsche 908

Porsche 911

Porsche 356

Dr. Ferry Porsche

# One Man's

My work and my hobby are the same: building cars that are fun to drive. Building cars that have the best possible engineering. Building *Porsches*.

The very first Porsche was the 356.

We had no marketing research to guide us. So it was designed to my personal specifications: small, lightweight, with good handling, and the power of a large car.

The idea was to build a racing car for the normal roads. And, in fact, in 1951, when the 356 was first entered at Le Mans, it won its class.

The second Porsche was, *and still is*, the 911.

First produced in 1964, it's forever young. Its engine displacement has been increased from the original 2 liters to 3 liters. And its output has been raised from 130 hp to 172 hp. Today, on the track, the 911 accelerates from 0 to 50 mph in 5.8 seconds. Its maximum speed: 139 mph.

The 911 and its derivatives have won countless major victories in motor sport, including rallies, hill climbs, and races. The 935 Turbo, for example, won the World Championship of Makes in 1976, '77, '78, and '79.

At Porsche, we view the race track as a proving ground. For on the track, under the stresses, surprises, and realities of competition, the best engineering wins. And what we learn from our race cars, we put into our production cars.

24

Porsche 928

Porsche 935 Turbo

Porsche 936 Turbo

Porsche 944

# Family.

The 917 Turbo Can-Am champion (1973) made turbo-charging practical for production cars.

The 936 Turbo Le Mans champion (1976,'77, '81) further advanced engine design and aerodynamic efficiency.

The 908 Targa Florio champion (1970) set the handling standards for all Porsches, including our new-generation 928.

The 928 has a 4.5-liter, 220-hp, aluminum Porsche V-8 engine. On the track, it accelerates from 0 to 50 mph in 6.0 seconds. And it has a maximum speed of 143 mph.

The 928 also has the Porsche transaxle design which produces balanced braking and improved cornering. And the 928 offers unprecedented comfort and luxury in a Porsche.

The newest Porsche is the 944.

It has a new 2.5-liter, 143-hp, aluminum Porsche engine, built at Zuffenhausen. On the track, it accelerates from 0 to 50 mph in 5.9 seconds. And it has a maximum speed of 130 mph.

The 944 also has the Porsche transaxle design, Porsche handling, and Porsche aerodynamics.

To me, the 944 is more than a new car. It is a new and true Porsche.

At Porsche, excellence is expected.

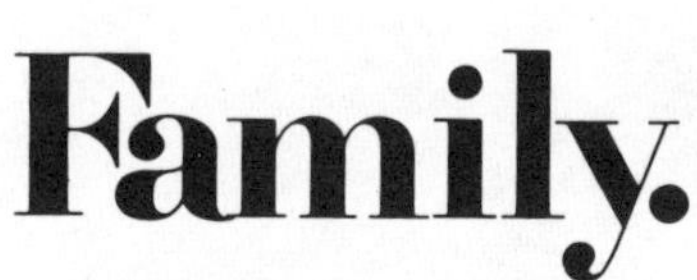

**PORSCHE + AUDI**

Dr. F. Porsche
Stuttgart

# 911SC Cabriolet

*Porsche's three-digit widget goes alfresco.*

• Sawing the tops off coupes was all the rage in Detroit during the summer of '82. When the chips had settled, convertibles could be had wearing six different nameplates—big showboat convertibles, compact convertibles, fast ones, slow ones, some with expensive price tags, others at low, low monthly payments. The Big Three's rush to roadsters took on religious overtones. It seemed that once enough roof panels were sacrificed to the gods, the sun would again shine on the Motor City.

Of course, the sun always shines in that happy corner of West Germany known as Zuffenhausen. The home of Porsche hasn't had to endure the two years of dreary sales that have plagued Motown; Zuffenhausen is the factory outlet of success. Engineering triumph follows engineering triumph, and customers vie for position in waiting lines for the latest three-digit widgets. As long as the model designation begins with the number "9," it's golden.

So why would Porsche take its oldest model, dip it in Detroit-think, and come up with a convertible? The answer is a man—an American, oddly enough—one Peter W. Schutz, who took over as chairman of the board of Porsche just two years ago. The 911SC convertible is the first major product for which Schutz can take the credit, so it is both a turning point for the firm and a portent of Porsches to come. As such, the Cabriolet is more phenomenon than fad: this is where Porsche shifts away from building pure engineering exercises and starts producing automobiles in response to legitimate market demands.

The 911 is one old crock of a car, not far from its twentieth birthday. Its branch of the Porsche family tree runs straight back to the second Porsche ever produced, the 1948 Type 356/2 (the first 356 was mid-engined). The 911 series has at one time or another con-

# 930S Turbo

*Let the men of Zuffenhausen build one for you.*

• Maybe you've been wondering what's so special about a Porsche. What you should do is drive this Porsche Turbo. Then you'll figure it out.

See, there's a special place in the heart that belongs to Porsche. That's because we expect a Porsche to be more than just an automobile. It's supposed to represent a way of thinking, a dedication to technological excellence. For Americans in particular, Porsche has become the focus of a nostalgic longing for standards of quality that seem to be missing from modern life.

This reverence for all things Porsche takes peculiar forms, like Porsche jackets, Porsche pillows, and Porsche sunglasses. It's also reflected in the pilgrimages true believers make to the Porsche factory and in the mystical quality attached to such relics of the trip as a visitor's pass or a sugar packet from the company dining room.

A far more reasonable embodiment of the Porsche spirit is this red 930 Tur-

bo that looks so much like a racing 935. It is literally a bespoke Porsche, a car custom-tailored by the foremost maker of sporting automobiles in the world to fit the vision of one driver.

Steven Knappenberger of Scottsdale Porsche + Audi (6925 East McDowell Road, Scottsdale, Arizona 85257; 602–941–0000) brought this car to America. Knappenberger, who decided the most appropriate outlet for his enthusiasm for Porsche's way of doing things would be a dealership, has made Scottsdale Porsche the leading Porsche store in Arizona. In addition to its regular trade, the store also imports one or two European-specification Porsches every month for U.S.-spec conversion; it was on a business trip to arrange the delivery of a European car that Knappenberger learned that Porsche had at last decided to get into the lucrative business of creating 935 look-alikes. Since he knows the difference between a conversion executed with tin snips and fi-

tained the fastest, the quickest-accelerating, the best-cornering, and the best-braking cars in America. With the demise of Checker, it is now the oldest car design in existence here.

Nonetheless, those who label the 911 a has-been are gradually being overruled by those who see truth, beauty, and a future in the car. Prof. Ernst Fuhrmann, the ex-chairman of Porsche, was embarrassed by the 911's crudity and prayed the 928 would replace it. Peter Schutz, to the contrary, considers the primeval Porsche "an exceptional technology carrier that will be a significant offering for an indefinite period of time." This man has casually mentioned composite body materials, a new, liquid-cooled engine, and four-wheel drive in the 911 context. If anything, the convertible is a conservative step into the future according to Schutz.

In truth, the 911SC is also a fantasy from Schutz's past. As the story goes, Schutz and Dr. Ferry Porsche were relaxing at the family villa one evening when Ferry explained that Porsche's system of product planning back in the early postwar years was simplicity itself: "We did no market research, no sales forecast, no return-on-investment analysis whatsoever. I built my dream car [the 356] and put it on sale." What a golden opportunity for Schutz to pipe

berglass and a conversion executed the Porsche way, Knappenberger placed his order.

The Porsche repair shop in Zuffenhausen, West Germany, actually does the work. In recent years it has turned from the repair of body damage and the restoration of old Porsches to customizing new cars. Now the repair shop will do almost anything, for a price; the only rule is that the modifications must be both durable and technically flawless. Those modifications now include new interior color and leather treatments, gearshift knobs in leather, wood, silver, gold, and even platinum, vanity cases, safes, stereos with up to 24 speakers, televisions, radio telephones, power steering, fittings for the physically handicapped, and many others. And since the fall of 1981, the repair shop also makes bodywork modifications.

This red Turbo went into the Porsche repair shop last March. It emerged in June. After being air-freighted to the U.S., it was sent to Alan Johnson Porsche + Audi in San Diego for federalizing. Just like the other Europeanspec cars Knappenberger imports, this Turbo is fully warrantied in all 50 states, with free towing and rental-car service. Scottsdale Porsche also stocks parts.

Yet this car is different from the other Porsches Knappenberger imports in

up with a pet dream of his own. "Let's build a 911 Cabriolet," he said to Ferry Porsche, and so it was.

Schutz had no way of knowing the whole world would be turning to convertibles as his Cabriolet came to market. Nor did he have any way of anticipating the magic he was mixing: the first time he ever drove a 911 was on the job at Porsche, and his comments were on the order of "Oh, my God." But since he started with a machine that is practically bursting with go, turn, and stop charisma, then tossed in the sights, sounds, and smells of nature, it comes as no surprise that the resulting entertainment level borders on overdose.

Technically, the exercise was very un-Porsche-like because it was so simple. The 911 Targa had been a mainstay for years, and the practicality of an open body with no roll bar had long ago been investigated and found entirely practical. Early in 1981, the old reports were dusted off and a convertible prototype was assembled over a two-month period. Durability tests proved that the Targa's basic structure would do the job. The Cabriolet debuted at the Frankfurt Auto Show that fall, proudly wearing its folded top and a venturesome four-wheel-drive bottom.

The top itself is one of those typically German overengineered devices; it ended up more metal than fabric. The forward third of the roof contains a large chunk of sheet steel to provide a stable base for the fabric and the latching hardware (two removable handles twist a pair of top hooks that engage sockets in the windshield header). Inside, the framework is hidden from view, except for one crossbow and a pair of fore-and-aft linkages in each rear corner. Concealed spring-loaded cables keep the

special ways. First, it might just be the ultimate Porsche, because of the personal relationship with the factory it implies. It's also the only one of its kind in the U.S. Finally, this look for the Turbo (itself scheduled for reintroduction to the U.S. by 1985, our spies tell us) could be the new shape for the 911.

Cosmetically, Porsche's slope-nosed conversion of the Turbo resembles many others, but the execution is different. The reason lies partially with Porsche's engineering standards, and partially with the fact that all such cars must be submitted to the TUV, an agency of the German government, before they can be offered for sale.

To begin with, the 935-type front and rear fenders and side valance panels are formed from corrosion-resistant, zinc-plated steel, just like standard Porsche bodywork. The factory offers either quad headlights integrated into the bumper or (Knappenberger's choice) pop-up 944 headlights; a servo motor concealed by a carpeted fiberglass blister in the trunk operates the headlights through an intricate linkage. The new front end features a central oil cooler and a fiberglass air dam. Hella driving lights replace the customary fog lights. The rear fenders conduct cooling air to the brakes through special vents. And since the side valance panels make it impossible to jack the car for a tire change in the usual manner, Porsche welded a fitting for a scissor jack beneath the belly pan and supplied a special, Japanese-made jack and a wooden adapter block.

At this point, Knappenberger and business partner Alan Johnson took over. Huge Pirelli P7 tires on wide BBS fifteen-inch wheels were installed (Porsche, fearful of aquaplaning and shortened wheel-bearing life, won't

outer fabric taut even at high speeds (we saw 136 mph with the top up, 127 mph with it folded). The back window is a flexible plastic affair that unzips on three sides with a large and durable-looking (but high-effort) zipper. A good share of the top's mechanism is made of aluminum alloy, so there isn't much of a weight penalty; Porsche says the Cabrio is 30 pounds lighter than the Targa, and we found it only 40 pounds heavier than the last coupe we tested.

Before we get to the joys of alfresco motoring, there are the negatives to consider. This is hardly one of those superb Italian convertible tops (like the Fiat and Alfa Spiders') that can be unlatched and flipped down over your shoulder during a pause at a stoplight. The toughest thing about lowering the Porsche's top is the fingernail-splitting rear-window zipper. The top itself is heavy, so it's far easier to stand outside the car to lever the assembly into the stowed position; the boot that's supplied (along with its own nifty, Velcro-sealed storage pouch) is literally a snap to install. The rough work comes when the first raindrops signal that fun in the sun is over: hefting the weighty roof into place is no trivial task. The tricky part is pulling the assembly down tightly enough against the windshield to engage the latches. A deep handhold is

provided on each side, and it helps if zipping the back window is saved for last; but even so, the effort really makes you long for two hands to tug and a third to twist. We did discover a shortcut (*not* recommended by the factory): when we let the front of the top drop against the windshield header from about a foot above it, we had far better luck engaging the latches.

Sealing, at least against water leaks,

make such a modification). A more radical 911 camshaft went into the engine, and a 934-type adjustable boost control went into the cockpit. The boost is set at 1.0 bar, instead of the 0.8 bar that Porsche recommends. Finally, a three-way catalyst (but without a feedback loop) and an EGR went into the exhaust system, and further ignition changes ensure the car meets air-emissions regs.

The 911 Turbo has been around a long time. Created originally to beat back BMW's expected entry into the sports-car market, the Turbo first appeared at the Frankfurt show in 1973. It reached production in 1975, and came to the U.S. in 1976 as the Turbo Carrera (only to disappear from here by 1979). But even though the Turbo's been around for a while, it still knocks your socks off.

From behind the wheel, the car feels vacuum-packed around you. You sit close to the front of the car to begin with, and the slope nose (a modification for aerodynamic stability developed by Porsche in 1975 to cope with the racing 935's unexpected top speed of 194 mph) makes you feel you're on top of the front bumper. Even with an unobstructed view of the road in front of you, however, the 325 horsepower in the engine compartment makes the windshield seem like the narrow slit of a bun-

ker as the speedometer needle begins its walk around the clock.

We decided extra running room would be necessary to let a performance car of this caliber off its leash, so we took ourselves deep into the wastes of the California desert to our exclusive top-speed proving ground. As the last glow of dusk disappeared, we set the 911 Cabriolet, headlights lit, next to the asphalt as a finish-line marker. The Turbo lined up 3.3 miles away.

You could see the Turbo's Hella

was flawless in our tester. Wind noise, however, reached the threshold of annoyance at about 80 mph. (Oddly enough, turbulence with the top down is quite tolerable at the same speed, partly because the front seats are relatively close to the windshield.) We found rough-surfaced roads agonizing at just about any speed, because they unleash the cavalcade of rattles, shakes, and squeaks that is part and parcel of *any* convertible. And sadly, Porsche has not advanced the state of the sun-visor art: the Cabrio's visors won't turn to the side, and the units in our tester wouldn't consistently stay in the up position. Lastly, there's the hazy back window. Our own L. Griffin reports that one officer of the law took advantage of the Cabriolet's restricted aft visibility to sneak up from behind and write a citation. It could happen to you. At least upgrades are under consideration, if European magazine reports are correct: both a rigid glass backlight and a power mechanism to take the drudgery out of

raising and lowering the roof.

All these petty matters drift into insignificance at the first blush of spring. In addition to the tousled hairstyle and the sunburned neck that distinguish convertible owners, you're treated to some of the finest sounds ever orchestrated by man or Mother Nature. Bend into a turn, and the front tires moan the message that understeer is present in 911s in manageable quantities these days. If you take your cornering *really* seriously, you'll find that manageable oversteer is also available just by lifting your right foot at the appropriate instant: the sound track shifts rearward, so it's the back rubber that sings the moody blues of tire slip. In the rhythm section, there's the ever-present whir of fan noise. At first it sounds like another Porsche hot on your tail, but you soon get accustomed to the mechanical melody. The rear-mounted orchestra is most inspiring during the upper half of its 6300-rpm rev range: with two cam chains, twelve rocker arms, dozens of cooling fins to resonate, and six ram-tuned induction throats on song, the Porsche symphony can be soul-stirring.

That's the magic Mr. Schutz has stumbled into. He took a car that's too great to kill and added a fresh dimension. All you Porsche fans who were wondering how long the 911 might last may now fret over other weighty matters. Like, for instance, when will the 911 get its two-liter, 200-horsepower, water-cooled V-8? —*Don Sherman*

---

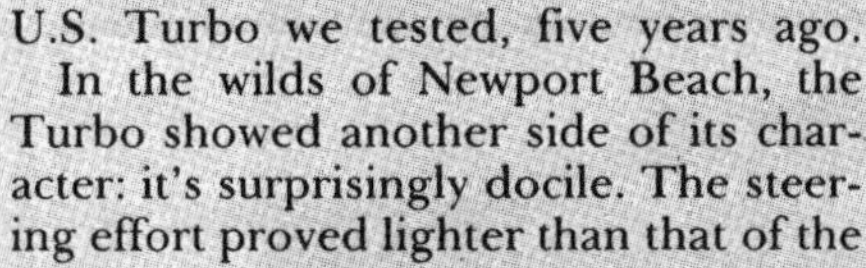

flame throwers approaching. Then you could hear the air crackling, the way it does after a clap of thunder. Finally, with the rush of a Southern Pacific freight like those running on the other side of the dry lake, the car hove into view, belly to the ground at 160 mph.

More urge lay underfoot, but the emissions calibration had produced substantial detonation; rather than invite piston meltdown, we backed off. As it turned out, the final performance specs closely resemble those of the last U.S. Turbo we tested, five years ago.

In the wilds of Newport Beach, the Turbo showed another side of its character: it's surprisingly docile. The steering effort proved lighter than that of the Cabriolet, as did the clutch and brake action. The four-speed transmission was also easy to use. The only tip-off to the wild heart of this beast was the heat waves shimmering off the rear deck.

The skidpad revealed a sizable discrepancy in left and right turning ability. We measured an eyeball-flattening 0.91 g to the left, but a disappointing 0.76 g to the right (for an average of 0.84 g). Even so, Alan Johnson, national SCCA driving champion several times in a 911, came back from a turn at the wheel with his eyes bright. "I know I'm going faster through a corner in a 928," he said, "but there's so much more for the driver to *do* with a rear-engined car. It's pure fun."

Indeed there's something special about this Porsche. It's not just another boutique car meant only for mounting the Newport Freeway. You could see this in the look of ecstasy that crossed the faces of the hard-core Porsche freaks who rapped on its fenders and discovered that Zuffenhausen steel lay under their knuckles, not City of Industry fiberglass. It was as if they had come face to face with a messenger from the Holy Land.

In some ways, the appeal of this unique Porsche defies explanation. It's essentially an old car, with vestiges of Porsche's first VW-powered sports car still in place. It's really good only in a straight line. Yet it has presence. It's the ultimate fantasy car, because it's built for the driver, not the passenger. And it meets the driver's fantasies with the highest standards of quality.

As it turns out, it isn't necessarily the way this Porsche performs that makes it so special; it's the way the car is engineered and built. It makes a statement about the way all cars should be built. And that's what a Porsche is all about. —*Michael Jordan*

# COUNTERPOINT

• Porsche 911s tend to divide the enthusiast population into two camps. Almost everyone agrees that 911s are outdated, cantankerous, overpriced Super Beetles. Thing is, either you succumb to their charm, or you don't. There's no middle ground. Kind of like love—you know?

I'll admit to falling on the side of the romantics, and so I find the new Cabriolet the most seductive 911 yet. The peak experience (as the Marin County hot-tub types like to call such events): a magnificent October afternoon, the autumnal color at its breathtaking zenith. A winding road. My significant other by my side.

Although the romantic me is an easy mark for the Cabriolet, the practical me says "no way." The Cab's rock-hard suspension threatens to turn it into a creaker before its time. The wind roar when the top's up is so serious at triple-digit speeds, I'd avoid taking it on long trips.

No, give me the 911 coupe. Cabriolets are the stuff of great love affairs. For happiness in a long-term relationship, I need something with a thicker skin.

*—Rich Ceppos*

The 911 Cabriolet leaves me with mixed emotions. I've always loved 911s, mostly because they're such sensible sports cars. On the sensible side, they're even-tempered, relatively quiet, and sumptuously comfortable; they require but moderate maintenance, and they don't cost an arm and a leg to operate. At the same time, they're absolutely killer sports cars, with as much enthusiast appeal as anything sold in America.

In the Cabriolet, this unique combination of practicality and pleasure has been seriously altered. The Cabriolet has squeaks and rattles. Hurricane-level wind noise takes the pleasure out of high-speed cruising. Even the coupe's excellent weather-tightness and visibility have been compromised by the Cabriolet's many seals, seams, and plastic rear window.

Of course, the Cabriolet offers full compensation for these degradations. On a beautiful day, with its top down, it's unbeatable. But its combination of sporting strong points and utilitarian weak points pushes the Cabriolet into the automotive-toy realm. That's hard for a 911 lover to take.

*—Csaba Csere*

Ten years ago, fools were saying the Porsche 911 was a dead duck, an over-the-hill anomaly of ill-advised design precepts that had little right to exist, let alone hold its head up, in the company of its so-called betters. But the 911 was fast, it was fun, it was satisfying, it was well built, and it was all these things in a way its so-called betters never came to understand. It was simply a complete thrill, and it still is.

Furthermore, folding the new cabrio top adds another vibrant thrill to those already on quickstep parade with the passage of every humming, whirring, snarling mile. It throttles your eardrums into submission, but warms your heart and whips you off through the heat and chill and sounds and smells of the real world, not a world once-removed by dint of safety glass and stamped roof panels. If you like the things I like, you will find yourself ignoring the shake in the cowl, the kick in the steering, the lumps in the ride, the shakes in the mirrors, the lack of air in the vents, and the murky images in the plastic window, and you'll wonder if all dead ducks are such fun.

*—Larry Griffin*

**Vehicle type:** rear-engine, rear-wheel-drive, 2-passenger, 2-door coupe

**Price as tested:** $95,000

**Sound system:** Blaupunkt 3001 AM/FM-stereo radio/cassette, 4 speakers

**ENGINE**
Type . . . . . . . . . . . turbocharged and intercooled flat 6, aluminum block and heads
Bore x stroke . . . . . . . . . 3.82 x 2.93 in, 97.0 x 74.4mm
Displacement . . . . . . . . . . . . 201 cu in, 3299cc
Compression ratio . . . . . . . . . . . . . . . . . . . . 7.0:1
Fuel system . . . . . . . . . . . . . . . . . Bosch K-Jetronic fuel injection
Emissions controls . . . . . . . . . 3-way catalytic converter, EGR, auxiliary air pump
Turbocharger . . . . . . . . . . . . . . . . . . . . . . . . . KKK
Waste gate . . . . . . . . . . . . . . . . . . . . . . . . . Porsche
Maximum boost pressure . . . . . . . . . . . . . . 14.5 psi
Valve gear . . . . . . . . . . . . chain-driven single overhead cam

Power (SAE net, estimated) . . . . . . . .325 bhp @ 5500 rpm
Torque (SAE net, estimated) . . . . . 335 lbs-ft @ 4000 rpm
Redline . . . . . . . . . . . . . . . . . . . . . . . . . . . . . 6700 rpm

**DRIVETRAIN**
Transmission . . . . . . . . . . . . . . . . . . . . . . . . . . 4-speed
Final-drive ratio . . . . . . . . . . . . . . . . . . 4.22:1, limited slip

| Gear | Ratio | Mph/1000 rpm | Max. test speed |
|---|---|---|---|
| I | 2.25 | 7.5 | 50 mph (6700 rpm) |
| II | 1.30 | 12.9 | 97 mph (6700 rpm) |
| III | 0.89 | 18.9 | 142 mph (6700 rpm) |
| IV | 0.63 | 26.7 | 160 mph (6700 rpm) |

**DIMENSIONS AND CAPACITIES**
Wheelbase . . . . . . . . . . . . . . . . . . . . . . . . . . . . 89.4 in
Track, F/R . . . . . . . . . . . . . . . . . . . . . . . 56.9/60.0 in
Length . . . . . . . . . . . . . . . . . . . . . . . . . . . . 168.9 in
Width . . . . . . . . . . . . . . . . . . . . . . . . . . . . . . 69.9 in
Height . . . . . . . . . . . . . . . . . . . . . . . . . . . . . 51.2 in
Ground clearance . . . . . . . . . . . . . . . . . . . . . . 4.2 in
Curb weight . . . . . . . . . . . . . . . . . . . . . . . . 2900 lbs
Weight distribution, F/R . . . . . . . . . . . . . . 36.0/64.0%
Fuel capacity . . . . . . . . . . . . . . . . . . . . . . . 21.1 gal
Oil capacity . . . . . . . . . . . . . . . . . . . . . . . . . 13.7 qt

**CHASSIS/BODY**
Type . . . . . . . . . . . . . . . . . . . . . . . . unit construction
Body material . . . . . . . . . . . . . . . . . .welded steel stampings

**SUSPENSION**
F: . . . . . . . ind, MacPherson strut, torsion bars, anti-sway bar
R: . . . . . . . ind, semi-trailing arm, torsion bars, anti-sway bar

**STEERING**
Type . . . . . . . . . . . . . . . . . . . . . . . . . . rack-and-pinion
Turns lock-to-lock . . . . . . . . . . . . . . . . . . . . . . . . . 3.0
Turning circle curb-to-curb . . . . . . . . . . . . . . . . . . 34.8 ft

**BRAKES**
F: . . . . . . . . . . . . . . . . . . . . . . 12.0 x 1.3-in vented disc
R: . . . . . . . . . . . . . . . . . . . . . . 12.0 x 1.1-in vented disc
Power assist . . . . . . . . . . . . . . . . . . . . . . . . . . .vacuum

**WHEELS AND TIRES**
Wheel size . . . . . . . . . . . . . . .F: 9.0 x 15 in; R: 11.0 x 15 in
Wheel type . . . . . . . . . . . . . . . . . . . . BBS modular aluminum
Tire make and size . . . . . . . . . . . . . . . Pirelli Cinturato P7,
F: 225/50VR-15; R: 345/35VR-15

# CAR AND DRIVER TEST RESULTS

| ACCELERATION | Seconds |
|---|---|
| Zero to 30 mph . . . . . . . . . . . . . . . . . . . . . . . . | 2.6 |
| 40 mph . . . . . . . . . . . . . . . . . . . . . . . . | 3.4 |
| 50 mph . . . . . . . . . . . . . . . . . . . . . . . . | 4.3 |
| 60 mph . . . . . . . . . . . . . . . . . . . . . . . . | 5.7 |
| 70 mph . . . . . . . . . . . . . . . . . . . . . . . . | 6.8 |
| 80 mph . . . . . . . . . . . . . . . . . . . . . . . . | 8.0 |
| 90 mph . . . . . . . . . . . . . . . . . . . . . . . . | 10.5 |
| 100 mph . . . . . . . . . . . . . . . . . . . . . . . . | 12.4 |
| 110 mph . . . . . . . . . . . . . . . . . . . . . . . . | 14.4 |
| Top-gear passing time, 30–50 mph . . . . . . . . . . . | 10.3 |
| 50–70 mph . . . . . . . . . | 9.6 |
| Standing ¼-mile . . . . . . . 13.8 sec @ 108 mph | |

Top speed . . . . . . . . . . . . . . . . . . . . . . . . . . . 160 mph

**BRAKING**
70–0 mph @ impending lockup . . . . . . . . . . . . . 165 ft
Modulation . . . . . . . . . . . . . . . . poor fair good **excellent**
Fade . . . . . . . . . . . . . . . . **none** moderate heavy
Front-rear balance . . . . . . . . . . . . . . poor fair **good**

**HANDLING**
Roadholding, 200-ft-dia skidpad . . . . . . . . . . 0.84 g
Understeer . . . . . . . . . . . . . minimal **moderate** excessive

**FUEL ECONOMY**
EPA city driving . . . . . . . . . . . . . . . . . . . . . . 14 mpg

**Vehicle type:** rear-engine, rear-wheel-drive, 2+2-passenger, 2-door convertible

**Price as tested:** $38,355

**Options on test car:** base 911SC Cabriolet, $34,450; forged aluminum wheels and Pirelli P7 tires, $1580; leather interior, $1000; Blaupunkt Monterey AM/FM-stereo radio/cassette, $795; alarm system, $200; sport shocks, $180; fog lights, $150.

**Sound system:** Blaupunkt Monterey SQR 22 AM/FM-stereo radio/cassette, 4 speakers

### ENGINE
Type ............... flat 6, aluminum block and heads
Bore x stroke .......... 3.74 x 2.77 in, 95.0 x 70.4mm
Displacement ..................... 183 cu in, 2994cc
Compression ratio ................................ 9.3:1
Fuel system ............. Bosch K-Jetronic fuel injection
Emissions controls ..... 3-way catalytic converter, feedback fuel-air-ratio control
Valve gear ............ chain-driven single overhead cam
Power (SAE net) ................. 172 bhp @ 5500 rpm
Torque (SAE net) ............ 189 lbs-ft @ 4200 rpm
Redline ................................. 6300 rpm

### DRIVETRAIN
Transmission ................................. 5-speed
Final-drive ratio ............. 3.88:1, limited slip

| Gear | Ratio | Mph/1000 rpm | Max. test speed |
| --- | --- | --- | --- |
| I | 3.18 | 5.9 | 37 mph (6300 rpm) |
| II | 1.83 | 10.2 | 64 mph (6300 rpm) |
| III | 1.26 | 14.8 | 93 mph (6300 rpm) |
| IV | 1.00 | 18.6 | 117 mph (6300 rpm) |
| V | 0.82 | 22.7 | 136 mph (6000 rpm) |

### DIMENSIONS AND CAPACITIES
Wheelbase ............................... 89.5 in
Track, F/R ......................... 53.6/52.8 in
Length ............................... 168.9 in
Width ................................ 65.0 in
Height ............................... 51.6 in
Curb weight ......................... 2740 lbs
Weight distribution, F/R ........... 38.0/62.0%
Fuel capacity ....................... 21.1 gal

### CHASSIS/BODY
Type ....................... unit construction
Body material ......... welded steel stampings

### INTERIOR
SAE volume, front seat ............. 43 cu ft
rear seat ................. 13 cu ft
trunk space ............... 4 cu ft
Front seats ........................... bucket
Recliner type ............. infinitely adjustable
General comfort ........... poor fair good **excellent**
Fore-and-aft support ....... poor fair good **excellent**
Lateral support ........... poor fair **good** excellent

### SUSPENSION
F: ...... ind, MacPherson strut, torsion bars, anti-sway bar
R: ....... ind, semi-trailing arm, torsion bars, anti-sway bar

### STEERING
Type .......................... rack-and-pinion
Turns lock-to-lock ....................... 3.1
Turning circle curb-to-curb .............. 34.0 ft

### BRAKES
F: ..................... 9.3 x 0.8-in vented disc
R: ..................... 9.6 x 0.8-in vented disc
Power assist ............................. vacuum

### WHEELS AND TIRES
Wheel size ....... F: 6.0 x 16 in; R: 7.0 x 16 in
Wheel type ...................... forged aluminum
Tire make and size ....... Pirelli Cinturato P7, F: 205/55VR-16; R: 225/50VR-16

## CAR AND DRIVER TEST RESULTS

### ACCELERATION
| | Seconds |
| --- | --- |
| Zero to 30 mph | 2.0 |
| 40 mph | 3.1 |
| 50 mph | 4.8 |
| 60 mph | 6.4 |
| 70 mph | 9.0 |
| 80 mph | 11.5 |
| 90 mph | 14.5 |
| 100 mph | 19.5 |
| 110 mph | 24.6 |
| Top-gear passing time, 30–50 mph | 8.3 |
| 50–70 mph | 9.1 |
| Standing ¼-mile | 14.8 sec @ 92 mph |
| Top speed | 136 mph |

### HANDLING
Roadholding, 282-ft-dia skidpad ............. 0.77 g
Understeer ............. **minimal** moderate excessive

### BRAKING
70–0 mph @ impending lockup .............. 175 ft

Modulation .................. poor fair **good** excellent
Fade ...................... **none** moderate heavy
Front-rear balance .................. poor fair **good**

### COAST-DOWN MEASUREMENTS
Road horsepower @ 50 mph .............. 13.0 hp
Friction and tire losses @ 50 mph ........... 5.0 hp
Aerodynamic drag @ 50 mph .............. 8.0 hp

### FUEL ECONOMY
EPA city driving ..................... **16 mpg**
EPA highway driving ................. 28 mpg
EPA combined driving ............... 20 mpg
C/D observed fuel economy ............ **22 mpg**

### INTERIOR SOUND LEVEL
Idle ............................... 63 dBA
Full-throttle acceleration .............. 87 dBA
70-mph cruising .................... 79 dBA
70-mph coasting ................... 77 dBA

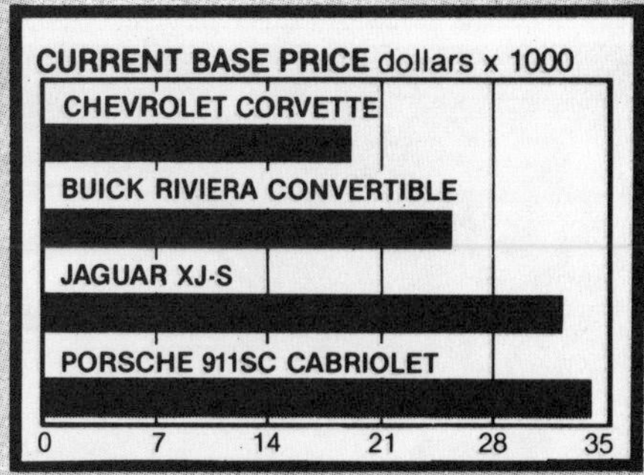


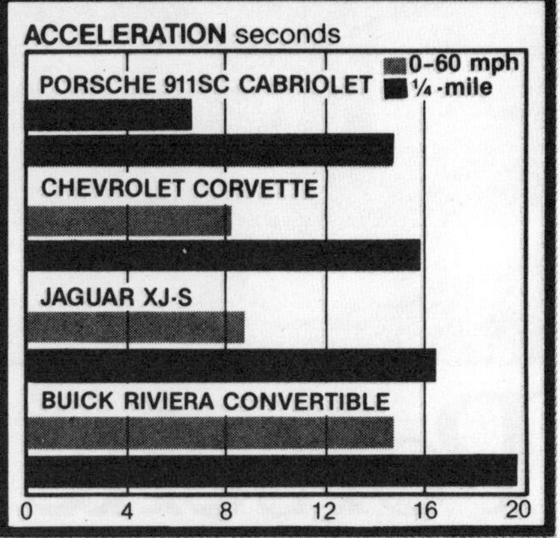


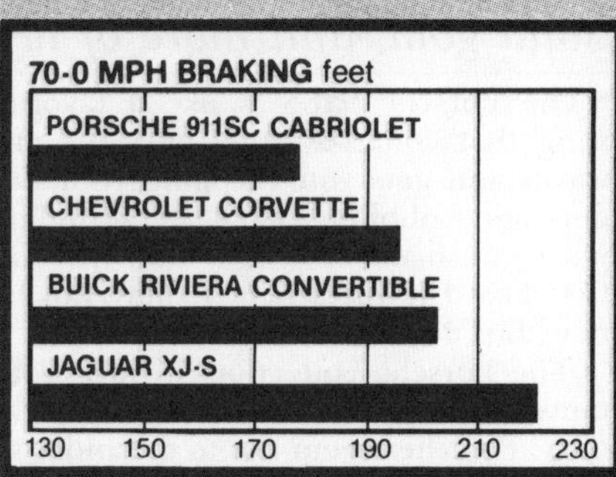


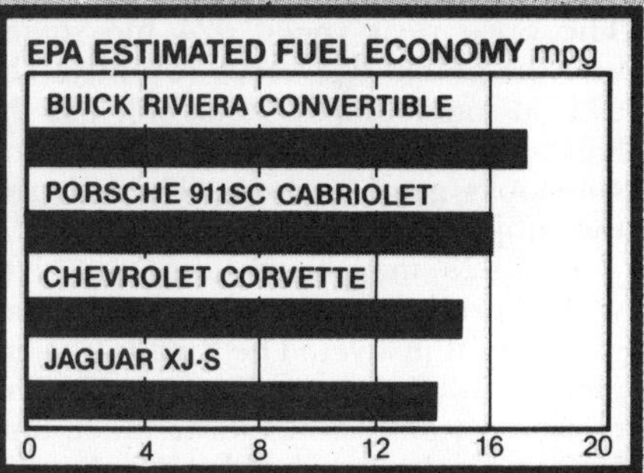


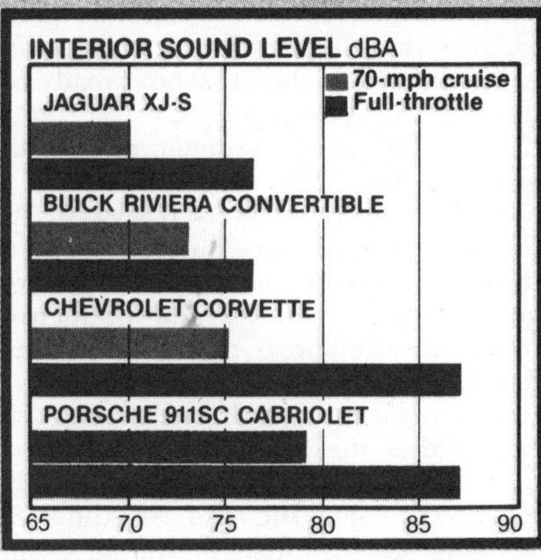

# Porsche 928S

## *Solid gold, and more of it.*

• The Porsche 928S is like a favorite song that comes wafting over the airwaves and gets you snapping your fingers and bobbing your head, sounding just as fresh and clear as it did when you first heard it years ago. It may not be new, but it's hard to believe it's old.

The Porsche collection, in fact, contains nothing but solid-gold hits these days; Porsches seem not to get older so much as better. Consider the evidence: The once weak-kneed 924 blossomed into the potent 944 series recently. The 911, at nineteen years old, still has the legs on just about all the new iron—and still looks great to boot. Then there's the subject of this report.

The 928 may be entering its sixth year of production, but don't get nostalgic about it just yet. The hard-bitten engineer-enthusiasts toiling at Zuffenhausen certainly don't waste much time lamenting the good old days. A good chunk of the reason Porsches are great drivers' cars is the firm's unswerving commitment to the slow but steady refinement of its designs.

This flow of development energy has given us the 928S, Porsche's idea of a new-and-improved, Mark II version. All of this year's 928 crop both here and in Europe receives the "S" designation and equipment, but the new model isn't really an attempt to redress any shortcomings of the original version. It didn't need much help. Last year's car was still at the top of the *C/D* hit parade, good enough to be considered the primo GT car in the land. No, the 928S is more like the next inevitable evolutionary step—produced because, well, it was *time*.

The wait was worth it. Every one of the revisions makes the 928S that much more mouthwatering. The big news is in the powertrain, where Porsche has accomplished a two-pronged improvement, increasing power and reducing fuel consumption at the same time.

Porsche started by boring out the 928's all-aluminum V-8 from 4474cc to 4664. A slight change in combustion-chamber design allowed a compression-ratio increase from 9.0:1 to 9.3 and—presto!—there are now 234 horses at your beck and call, fourteen more than before. (European models, unfettered by our strict emissions controls and low-octane fuel, have a 10.0:1 compression ratio and deliver 288 hp.)

To squeeze a bit more mileage from each tank, the final-drive ratio was dropped from 2.75:1 to 2.27. The gearbox's bottom four ratios were lowered a bit and spread out a touch farther to

compensate, but fifth gear is unchanged, so the engine loafs along at only 3000 rpm at 100 mph in top gear. EPA city mileage remains at 16 mpg, but highway mileage has improved from 25 mpg to 27.

Porsche combined the new mechanicals with what was previously referred to as the competition-package option to create the American S. The goodies include a chin spoiler, a deck-lid spoiler, and flat-faced alloy wheels, all of which combine to drop the drag coefficient of the bulbous body from 0.41 to a more respectable 0.38. A set of 225/50VR-16 Pirelli P7 radials replaces the 215/60VR-15 tires that were formerly standard. Larger brakes and sport shocks shore up the chassis. Leather seats, air, cruise, central locking, power windows, and more are also part of the deal. The only other significant change—as you might expect, given the increased standard-equipment list—is the price. It's up to $43,000, from $39,500.

It might seem hard to believe, but the feeling around these parts is that the 928S actually gives you your money's worth in today's inflated market. Here is one car that can do it all, friends, and never breathe hard.

You want numbers? Try zero-to-sixty in 6.2 seconds (a 0.6-second improvement from last year). Quarter-miles zip by in just 14.7 seconds at 94 mph (quicker than before by 0.4 second and 3 mph). Top speed is up to a blazing 144 mph, a dividend of 9 mph, making the 928S the fastest car sold in America. Just for good measure, it stops from 70 mph in an impressive 188 feet and puts an 0.80-g hold on the road.

But none of this adequately depicts life with 928. Numbers can't capture this car's double-agent personality: a killer instinct coupled with luxocar civility and the kind of bulletproof solidity you'd normally associate with a Mercedes.

Aside from its stiff-legged ride, the 928S can be as docile as an Eldorado when you troll down to the cleaners. Call for speed, though, and the needle heads for 120 mph in an effortless gush. It's all so easy you have to force yourself to put both hands on the wheel at triple-digit speeds. And on the back roads, there's more magic than most folks will ever know what to do with.

Flights of hyperbole over the 928 are nothing new. This car has been a bullet on the charts from day one; this year, it's all just a little sweeter. Even the once-balky shift linkage running to the rear transaxle has been tamed. This underscores just what makes the 928S, and all Porsches, so special: in each replay of the original, there's always a little bit more gold.

*—Rich Ceppos*

---

**Vehicle type:** front-engine, rear-wheel-drive, 2+2-passenger, 3-door coupe
**Price as tested:** $43,940 (base price: $43,000)
Engine type: V-8, aluminum block and heads, Bosch L-Jetronic fuel injection

| | |
|---|---|
| Displacement | 285 cu in, 4664cc |
| Power (SAE net) | 234 bhp @ 5500 rpm |
| Transmission | 5-speed |
| Wheelbase | 98.4 in |
| Length | 175.7 in |
| Curb weight | 3360 lbs |
| Zero to 60 mph | 6.2 sec |
| Zero to 100 mph | 17.8 sec |
| Standing ¼-mile | 14.7 sec @ 94 mph |
| Top speed | 144 mph |
| Braking, 70–0 mph | 188 ft |
| Roadholding, 282-ft-dia skidpad | 0.80 g |
| EPA fuel economy, city driving | 16 mpg |
| *C/D* observed fuel economy | 13 mpg |

# Porsche 944 Automatic

*Minus some of the magic.*

• If your significant other can't comprehend your feeling of lust toward the 944, come to us. We think lust is a perfectly acceptable emotion. When your dreamboat is a 944, that six-hour jaunt to the Poconos becomes the honeymoon. The long crosstown commute at the close of a grueling workday seems reward enough for your toil. The Monterey peninsula isn't all that far from Cape Cod anymore, is it?

Having a 944 in the *Car and Driver* motor pool never fails to add new meaning to the road testers' lives. And it's guaranteed to elicit a rare display of rank-pulling by our senior staff members when the time comes to divvy up the fleet for the weekend. The 944 is just that kind of car.

But a funny thing happens to all that magic when Porsche puts an automatic transmission into the 944. It kind of goes away. Please don't misunderstand: the 944 automatic remains on our list of favorite cars, but lust no longer applies. "Intense like" is more appropriate.

All the fine qualities that contribute to greatness in the 944 are still there. The all-aluminum 2.5-liter fuel-injected four-cylinder engine with the Mitsubishi-style counterrotating balance shafts is still as quiet at high revs as a six- or eight-cylinder engine. And it still packs a healthy 143-horsepower wallop at 5500 rpm, though the automatic saps more of its energy than the five-speed does (1.9 seconds slower from 0 to 60 mph, 1.1 seconds slower in the standing quarter-mile).

Nothing has been changed in the four-wheel independent suspension system; the 944's ride is exceptionally stable for a sports car. It revels in a good, clean piece of roadway, yet doesn't fuss or fret when the going gets rough. The 944's unassisted rack-and-pinion steering is among the best we've come across, and the car's excellent handling doesn't suffer a lick from the addition of an automatic. The transmission is still located in the rear of the car (as is the torque converter); the engine is still in the front. The only difference in the automatic version is that fluid lines had to be strung far forward to the transmission cooler, an integral part of the radiator. This unusual drivetrain layout gives the 944 an almost dead-on fifty-fifty weight distribution, which some say is the ultimate. (*C/D*'s technoids disagree, however.)

Our disappointment with the automatic isn't based on anything as concrete as engineering. The VW-and-Audi-made 010 automatic is a fine

transmission. It is, in fact, the same gearbox used behind both four- and five-cylinder engines in various VWs and Audis. (The 928's four-speed automatic is built by Mercedes-Benz.) Of course, each application has its own specific valve body, but the basic design of the three-speed 010 is flexible enough to accommodate everything from a meek four-cylinder right up to the 170-hp European-spec Audi 5000 turbo five-cylinder.

The 944's fat torque curve is a natural for an automatic transmission. By 2000 rpm, the engine is producing 93 percent of its 137 pounds-feet of peak torque, and from 2500 to 5500 rpm the curve is effectively flat. A final-drive ratio of 3.45:1 ensures some punch in the low range, while wide-open-throttle upshifts occur at 6000 rpm from first to second, and at 6200 rpm from second to third.

The 944's allure goes beyond the inner workings to the mean look of its exaggerated fender flares and arresting yellow instrument graphics. Body-hugging sport seats and the pleasing shape and heft of the leather-clad steering wheel and shifter knob also add to the 944's vitality. It's a marvelous car. So what *is* our complaint?

The Porsche 944 is tailor-made for those who drive for a hobby. It begs to be controlled, to be driven into the ground. When some of the control is robbed from the driver—poof!—some of the feeling goes with it. And the E.T.-style shift knob is not a satisfactory substitute.

There are those who are intimidated by Porsche's performance image, and others who view a clutch pedal as an obstacle. The 944 automatic has enough going for it to keep those unfortunates happy; they'll never know what they're missing.

But if you truly lust after a good turn in a hot driver's car, don't bug your dealer for a test run in the automatic. Stick with the five-speed.

—*Jean Lindamood*

**Vehicle type:** front-engine, rear-wheel-drive, 2+2-passenger, 3-door coupe
**Price as tested:** $22,150 (base price: $18,980)
Engine type: 4-in-line, aluminum block and head, Bosch L-Jetronic fuel injection

| | |
|---|---|
| Displacement | 151 cu in, 2479cc |
| Power (SAE net) | 143 bhp @ 5500 rpm |
| Transmission | 3-speed automatic |
| Wheelbase | 94.5 in |
| Length | 170.0 in |
| Curb weight | 2900 lbs |
| Zero to 60 mph | 9.4 sec |
| Zero to 100 mph | 29.2 sec |
| Standing ¼-mile | 16.8 sec @ 84 mph |
| Top speed | 130 mph |
| Braking, 70–0 mph | 183 ft |
| Roadholding, 282-ft-dia skidpad | 0.81 g |
| Road horsepower @ 50 mph | 15.5 hp |
| EPA fuel economy, city driving | 21 mpg |
| *C/D* observed fuel economy | 17 mpg |

PHOTOGRAPHY AARON KILEY

# DP 935

*Fantasies fulfilled, no waiting.*

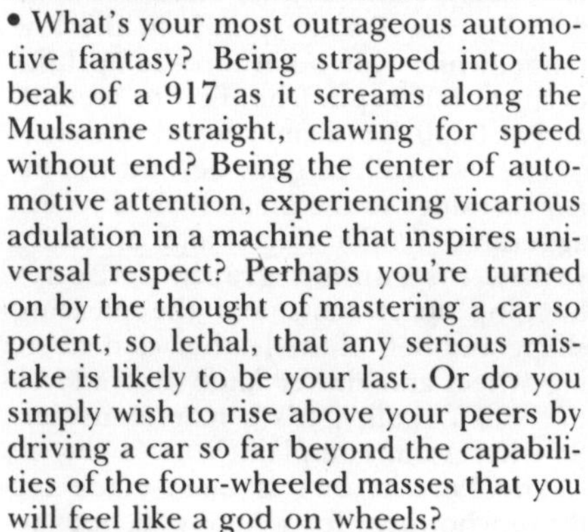

• What's your most outrageous automotive fantasy? Being strapped into the beak of a 917 as it screams along the Mulsanne straight, clawing for speed without end? Being the center of automotive attention, experiencing vicarious adulation in a machine that inspires universal respect? Perhaps you're turned on by the thought of mastering a car so potent, so lethal, that any serious mistake is likely to be your last. Or do you simply wish to rise above your peers by driving a car so far beyond the capabilities of the four-wheeled masses that you will feel like a god on wheels?

Here's a car that can fulfill every one of these fantasies with startling realism. The DP 935 started life as a Porsche 930, but has been transformed into a roadgoing facsimile of a racing 935. The fiberglass panels that accomplish this were flawlessly executed by Designer Plastics Automobilbau, the bodywork couturier of Kremer Racing, among others. The oversize rear spoiler; the hopelessly low, curb-fouling, deeply ducted front air dam; the tiny, aerodynamic, nearly useless mirrors; the running boards; and the broad, multifariously vented fenders combine to create the 935 look convincingly enough to freeze knowledgeable car lovers dead in their tracks.

Inside the DP 935, the competition imagery continues. Although the sumptuous Porsche leather interior is retained, a serious go-fast ambiance is imparted by the oversize boost gauge; the pure-racing, perforated-aluminum dead pedal; and the very businesslike adjustable boost knob directly astern of the shift lever. A pair of Porsche sport seats are comfortable and supportive enough to locate the passengers against the omnidirectional accelerations this car produces. No sound system is provided, or particularly missed: sensory overload is already ensured by the DP's raspy exhaust, over-the-nose-cone view, and heavy thrust.

Generating this thrust is a DP modified engine. The carefully assembled stock 930 motor was ripped apart and blueprinted to stricter-than-Zuffenhausen standards, the heads were ported and polished (along with the turbo's inlet housing), and the compression was bumped up a tad. Peak boost is set at 16 psi, more than 3 psi higher than in a stock 930. The engine was also emissions-certified in America with a catalyst in lieu of the muffler; this modification substituted a sharper rasp, though not much louder, for the standard 930's whooshy sound. It did not, however, permit the engine to run on any unleaded fuel that we could find, so we resorted to heavy transfusions of octane booster to keep detonation at bay.

Our miscreancy was rewarded with rocket-sled performance. The DP 935 lunged from a standing start to 60 mph in a mere 4.6 seconds, 100 came up in but 10.5 seconds, and 130 was only another 8.8 seconds away. It scorched the quarter-mile in 12.8 seconds at 112 mph. But even these fabulous figures don't convey the car's sheer thrust. The DP 935 delivers an incredibly strong shove in the back, and the punch never seems to quit. Even at 120 mph it pulls harder than a Scirocco at 50. Not until well past 140 mph does your breath come back and the acceleration of this juggernaut fall to a familiar level.

In light of this tremendous proclivity for speed, the car's peak of 162 mph is a bit disappointing. But with the DP's fifteen-inch tires (the stock diameter is sixteen inches), the engine is revving well beyond its power peak at top speed. The exhaust-plugging, aftermarket catalyst probably didn't help the power curve past 6000 rpm either.

The smaller-diameter tires are part of the DP 935's chassis meliorations. Porsche forged wheels, eight and eleven inches wide, front and rear, are shod with 205/50VR-15 and 285/40VR-15 Pirelli P7 gumballs. The suspension is lowered and stiffened as well. These changes firm up the ride considerably, but the motion left within the limited suspension travel is fluid and supple. Like the rest of the car, this suspension feels best in triple-digit speed ranges. It also happens to work around corners: on the skidpad, we measured 0.84 g of stick, the best we've ever seen on a roadgoing Porsche.

These landmark performance measurements bolster the DP 935's position as a paragon of automotive omnipotence. And for a mere $95,000, Richard Buxbaum of Classic Motors in Chicago can make it yours. That may seem a bit high, but fantasy fulfillment never comes cheap.

—*Csaba Csere*

**Vehicle type:** rear-engine, rear-wheel-drive, 2-passenger, 2-door coupe
**Price as tested:** $95,000

| | |
|---|---|
| Engine type: turbocharged and intercooled flat 6, aluminum block and heads, Bosch K-Jetronic fuel injection | |
| Displacement | 201 cu in, 3299cc |
| Power (SAE net, estimated) | 370 bhp @ 5500 rpm |
| Transmission | 4-speed |
| Wheelbase | 89.4 in |
| Length | 170.0 in |
| Curb weight | 2910 lbs |
| Zero to 60 mph | 4.6 sec |
| Zero to 100 mph | 10.5 sec |
| Zero to 130 mph | 19.3 sec |
| Standing ¼-mile | 12.8 sec @ 112 mph |
| Top speed | 162 mph |
| Braking, 70–0 mph | 179 ft |
| Roadholding, 282-ft-dia skidpad | 0.84 g |
| Road horsepower @ 50 mph | 13.5 hp |
| *C/D* observed fuel economy | 17 mpg |

# PORSCHE 911 CARRERA

Like Bob Hope, the Porsche 911 seems to have always been here and appears to get better every year. Not only does it get the Carrera name now, but it also gets a new fuel-injected, 3.2-liter engine with 28 more horsepower than last year's. For added thrills, 930 Turbo bodywork and chassis are available as a factory option.

Porsches have always given good kinesthetics, and this year is no exception. The raspy, air-cooled 3.2-liter flat six makes enough muscle to shove you in the back when you push the yes pedal. The tires hang on tenaciously. And the brakes are plain awesome.

Despite years of development, the Carrera's character is still intact. When you hurry, it feels as if there's an unruly beast inside trying to break out. More than one neophyte owner has felt the Carrera's tail-happy sting when making like Mario Andretti. But master the Carrera, and you have conquered a truly formidable automobile. It's an accomplishment worth the effort, and one of the things that make the Carrera a savory experience.

**Importer:** Porsche + Audi Division
Volkswagen of America, Inc.
Troy, Michigan 48099

**Base price:** $31,950–36,450
**Vehicle type:** rear-engine, rear-drive
**Body styles:** 2-door convertible, 2-door coupe, 2-door targa

**DIMENSIONS**

| | |
|---|---|
| Wheelbase | 89.4 in |
| Track, F/R | 53.6/52.8 in |
| Length | 168.9 in |
| Width | 65.0 in |
| Height | 52.0 in |
| Curb weight | 2750 lbs |
| Fuel-tank capacity | 21.1 gal |

**ENGINE**
SOHC 3.2-liter flat 6

**TRANSMISSION**
5-sp man

**SUSPENSION**
F: ind, MacPherson strut, torsion bars
R: ind, torsion bars

**BRAKES**
F: vented disc
R: vented disc

**EPA ESTIMATED FUEL ECONOMY**
City 20 mpg

# Porsche 911 Carrera

*Germany's holy roller, the fastest factory-fresh,
finger-in-the-socket, load-and-cock-it full-time pocket rocket. Amen.*

• It is the evil weevil, the rock-solid, steely-eyed grim reaper of sporting cars, the paragon of knife-edged incisiveness and buttoned-down insanity. More than any other factory-fresh passenger car available here today, the Porsche 911 Carrera is the absolute embodiment of clench-jawed, tight-fisted, slit-eyed enthusiasm run amok, a car for making the landscape pass with explosive fluidity. Strange that a car so serious can bring such unadulterated joy, but there you are, sporting an enormous, cheek-splitting leer when you unstrap and step out. You devil, you.

The former 911SC thrilled the hardcore, adrenalin-addicted, pop-eyed performance toadies among us, offering a truly startling mix of performance and practical-

ity, but it is no more. The Carrera replaces the 911SC, and it embodies all the same simultaneously outrageous and sensible qualities and more. The 911SC was fast, but the Carrera is a bullet. Firing from 0 to 60 mph in 5.3 seconds, an improvement of more than a full second, "shot out of a gun" covers it. And, although the 911SC returned a resolute 16 mpg on the EPA city cycle, the Carrera offers no less than a whopping 20 mpg.

Livability has even been improved somewhat in Porsche's sporting but anachronistic interior. The car also goes down the road better, feeling more civil and less likely to bolt—all due to the Porsche penchant for constant improvement. The term "running changes" carries a double meaning

in Porsche's corner of West Germany.

This tough act to follow is located mostly under the rear deck lid, which carries the inspirational Carrera nameplate, earlier versions also having possessed well-deserved reputations for rocketry. Porsche offers no fewer than three choices of shell configurations for the 911 Carrera: coupe, Targa, and Cabriolet. In response to the growing hue and cry among its customers for greater individuality, the fully enclosed shell is available not only with the same relatively subdued lower bodywork worn by its airier brothers (which share a new front spoiler housing the fog lamps), but also with a larger air dam and the most instantly recognizable passenger-car aerodynamic device of all time, the Stuttgart whale tail.

And, finally, the killer European Turbo treatment, meaning full-house bodywork and running gear, is available for everything but the engine.

Given the Carrera's rousing performance, the current lack of a turbocharged offering is not really very disappointing, particularly in light of the accompanying fuel-economy boost. Newly stuffed into the 911's rump are two additional tenths of a liter of displacement, bumping the single-overhead-cam flat six to 3.2 liters. At 95.0 millimeters, its bore remains the same, but its stroke has been elongated from 70.4 to 74.4 millimeters, matching the Turbo engine's. Although Porsche is perpetually searching for improved mechanical efficiency, it is nonetheless a surprise that the Carrera's engine is 80 percent new.

Reshaped pistons and combustion chambers within the air-cooled aluminum block and heads bring a modest compression-ratio increase, from 9.3:1 to 9.5:1. New tuned intake manifolding and Bosch Motronic fuel injection (very similar in principle to Porsche's version on the four-cylinder 944) are fitted. The all-electronic Motronic system is replacing Bosch's well-known mechanical K-Jetronic injection for a number of manufacturers because it produces extremely accurate operational tolerances for fuel mixture and ignition timing, prime contributors to efficiency. An overrun fuel shut-off minimizes unnecessary waste when the throttle is closed, al-

though this may have been the cause of the part-throttle on-off power surges we noticed during testing. The new idle-speed regulator that is a part of the Motronic system also had a hard time keeping the engine running stably at no load, particularly when cold.

When it comes to pure "go," however, the changes have pumped up an already healthy 172 bhp at 5500 rpm to a righteously distended 200 bhp at 5900 rpm. Maximum torque has risen from 175 pounds-feet at 4200 rpm to 185 pounds-feet at a lofty 4800 rpm, and the growling, thrumming flat six remains in the forefront of the world's grunt-and-git, instant-forward rushers.

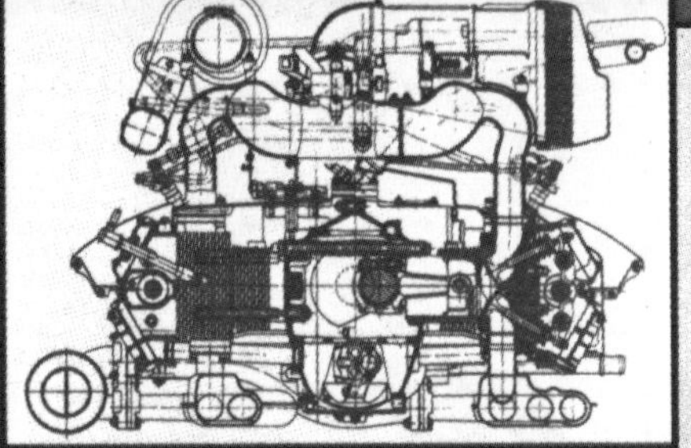

The world's only air-cooled flat six is back for 1984 with 200 horsepower, thanks to a longer stroke and new fuel injection.

One of the reasons, of course, for the Carrera's monumental acceleration capability is its out-back engine, which presses the 225/50VR-16 Goodyear NCTs into the pavement with all the vengeance that almost 60 percent of the Carrera's 2760 pounds can provide. Handling lesser duties up front, 205/55VR-16 NCTs are fitted to 6.0-inch-wide alloys, which are an inch narrower than those at the rear.

· The inequality of the Carrera's front and rear footprints shows that Porsche is still heavily involved in offsetting the rear-engine layout's inherent behavioral imbalance (read, strong trailing-throttle oversteer). The Carrera can and does demonstrate a will of its own, but its big-tired sense of right and wrong has produced a pretty remarkable taming of a once thoroughly unpredictable device.

The shifter seems paradoxically notchy and vague to some drivers, but familiarity smooths its potential glitches. Half the *C/D* staff loved it at first touch, the other half begrudgingly accepted it later. Nobody has any such problem with the brakes. The Carrera's four-wheel discs and rear weight bias have produced over the years what we roundly consider the best passenger-car braking to be found. For 1984, virtually the whole stopping system has been overhauled to handle the newfound horsepower. Front and rear rotors are seventeen percent thicker, there are wider-mouthed calipers to match, the vacuum booster is one inch larger in diameter, and a brake-effort proportioning valve has been added. This last item is particularly effective because it has allowed the engineers to reset the system's basic

balance. Low-speed, low-coefficient-of-friction front lockup has been a problem for years in the 911, but finally, Porsche has adjusted the balance toward the rear brakes. To keep them from locking prematurely during high-g, good-traction stops, the proportioning valve comes into play, automatically limiting rear brake-line pressure, shifting the balance back toward the front brakes. In our 70-to-0 stopping tests, we found that the right rear wheel was prone to lock first but that there was no loss of stability. Braking distances, at 184 feet, were as short as those of any other car available in America.

The steering garners good marks for precision and speed of response. It loses points for stout kickback over bumps, though this Carrera's wheel palsy did feel a bit more damped in our hands than the old 911's. There is less wriggle than before and cleaner tracking. The independent suspension is still all firm resilience, and even without the add-on aero aids, the Carrera sports noticeably improved high-speed stability and resistance to crosswinds. With the aids, stability is further improved. The really wide Turbo bodywork would undoubtedly cost some top speed.

Despite claimed improvements, the climatological outlook in the Carrera is something of a joke in this day and age. The airflow system is hard to understand, slow to respond, and riddled with gaps in coverage. The windshield-wiper control stalk of-

fers three speeds (including one fast enough to maim slow-moving pump jockeys); a separate dash knoblet must be activated for intermittent wiping. Instrumentation is one of the Carrera's long suits. It is formally dressed—everything in black and white—clean, handsome, and very complete; this part of the ergonomics has been a high priority at Porsche for many years. The most recently added fittings, however, have been tacked here and there wherever their plumbing would fit, the result being a hassle of ill-marked, often concealed impedimenta. Not even such standard accouterments as air, leather seats, and power windows take the edge off the interior's confusion beneath its businesslike façade. The optional automatic heat control and electrically heatable windshield would help (the two-position rear-defogger switch works well enough to get a barbecue going, but the front defroster is for laughs). Cruise control, an electric sunroof, and a factory-installed burglar-alarm system are some consolation, and the widely adjustable electronic Blaupunkt AM/FM/cassette stereo issues forth good sound.

But those who buy a Carrera because they looked at its features instead of actually driving one of these headlong, head-strong, drum-tight motorcars may be in for a big shock. And what could be finer?

Here is a car that humps from nothing to 30 mph in 1.9 seconds, to 100 in just 13.9 seconds (exactly a quarter-mile's worth), to

## COUNTERPOINT

• Despite the recent resurgence of performance cars, most manufacturers are still loath to bring really serious speed to market in America. BMW has kept its best motors at home for years. General Motors just canceled the turbo Fiero because it was too fast. Mercedes, though shamed into importing its big-block V-8, has seriously emasculated it. They all bray the same lame excuses— the 55-mph speed limit, emissions and CAFE regulations, even the unfavorable legal climate. Soon they'll be blaming sunspots.

Now along comes Porsche with the new Carrera, blowing them all out of the water—with a twenty-year-old design, no less. It's a car that can do accelerative battle with the seven-liter muscle cars that were popular in its infancy, yet it still meets 1984 emissions regulations. It's a car that can cover 2.5 miles per minute yet achieves 20 mpg on the EPA's city cycle. A car that coddles two adults, handles capably, stops on a dime, and is built like a tank. I'm sorry, gentlemen, but your excuses just don't cut it anymore.          —*Csaba Csere*

I just came in from driving the Carrera. Should I go and kiss the publisher's feet in thanks for hiring me, or should I throw myself on the floor and have a tantrum, knowing that I'll probably never be able to afford one?

If you're looking for an objective comment from me about this car, I'll tell you that the heater controls are medieval and that the shifter is heavy in city traffic. Don't push me beyond that; this is my car. The optional sport seat is perfect and the driving position is excellent, both imperative for the hard driving that the Carrera prefers. You couldn't ask for a better set of performance gauges; the brakes are well balanced, linear, and exceptionally effective; and the new engine feels spectacular.

I want this car.

Now back to my financial situation. I can run a pretty clean bead, I do windows, and I am very good with children. Available on an occasional Saturday. Write in care of this magazine.
—*Jean Lindamood*

In Germany, they say the old girl has another ten or twenty years of life in her, and I sincerely hope that's true. The Carrera is easy to love largely because it's so hard to drive. Decoding the heater is a chore. Your left hand must be trained to operate the ignition key. The gas- and brake-pedal locations make heeling-and-toeing a ten on the index of difficulty. The steering kicks in your hands, the shifter's rubbery, and it's taboo to lift off the gas at the wrong time.

But really, who said running up to this car's limits had to be easy? The harder it is to do, the better it feels when you get it right, right? So over the years the 911 has become the professional driver's car. Every time you start the engine, it screams, "Don't try me unless you're good." A trip to the 7-Eleven is a challenge, and every smoothly executed sweeper or well-coordinated run through the gears is its own reward. You may practice your Fangio moves all you want on a Toyota; but to perfect your craft behind the wheel, nothing beats a 911.          —*Don Sherman*

130 in 29.7 seconds, and very quickly up to a true, redline-limited terminal velocity of 149 mph—all of which makes entire parking lots' worth of formerly all-heroic performance luminaries look positively puny. The Carrera is simply the fastest factory-offered car for sale in the United States, no contest. (The Lamborghini factory is not yet delivering Countaches, as far as we know.) In addition, the Carrera is a Porsche. Surveying the world from its snug-fitting sport seats and feeling its every powerful pulse of communication, you'll feel that the Carrera possesses indestructibility.

Once upon a time, not so long ago, 911s were beginning to feel like somebody's idea of a bad joke that had run much too long in the telling. Over the past five years, Porsche has turned the tables, girding the cars' loins further still for extremely heavy duties in the lists of passionate motoring.

The Carrera is a car to get down and wrestle with. In exchange you will come away winded, exhilarated, and probably laughing aloud, sure of why it was that you first came to love the evil weevil, and sure that you still do.          —*Larry Griffin*

**Vehicle type:** rear-engine, rear-wheel-drive, 2+2-passenger, 2-door coupe

**Price as tested:** $37,075

**Options on test car:** base Porsche 911 Carrera, $31,950; 16-inch wheels and tires, $1580; front and rear spoilers, $1325; electric sunroof, $940; AM/FM-stereo radio/cassette, $600; cruise control, $320; sport seats, $250; black headliner, $70; extended steering-wheel hub, $40.

**Sound system:** Blaupunkt Monterey AM/FM-stereo radio/cassette, 4 speakers, 8 watts per channel

### ENGINE
Type ................... flat 6, aluminum block and heads
Bore x stroke .......... 3.74 x 2.93 in, 95.0 x 74.4mm
Displacement ........................ 193 cu in, 3164cc
Compression ratio ................................ 9.5:1
Engine-control system ............... Bosch Motronic
Emissions controls .....3-way catalytic converter, feedback fuel-air-ratio control
Valve gear ....... chain-driven single overhead cam
Power (SAE net) ............... 200 bhp @ 5900 rpm
Torque (SAE net) .......... 185 lbs-ft @ 4800 rpm
Redline ............................................ 6300 rpm

### DRIVETRAIN
Transmission............................... 5-speed
Final-drive ratio ................................ 3.88:1

| Gear | Ratio | Mph/1000 rpm | Max. test speed |
|------|-------|--------------|-----------------|
| I | 3.18 | 5.8 | 37 mph (6300 rpm) |
| II | 1.78 | 10.4 | 66 mph (6300 rpm) |
| III | 1.26 | 14.7 | 93 mph (6300 rpm) |
| IV | 1.00 | 18.5 | 117 mph (6300 rpm) |
| V | 0.79 | 23.6 | 149 mph (6300 rpm) |

### DIMENSIONS AND CAPACITIES
Wheelbase ......................................... 89.4 in
Track, F/R ................................... 54.0/54.3 in
Length ........................................... 168.9 in
Width ............................................. 65.0 in
Height ............................................ 52.0 in
Frontal area ................................. 19.1 sq ft
Curb weight ................................... 2760 lbs
Weight distribution, F/R .................. 41.3/58.7 %

### CHASSIS/BODY
Type ................................. unit construction
Body material .................. welded steel stampings

### INTERIOR
SAE volume, front seat ......................... 43 cu ft
rear seat .......................... 13 cu ft
trunk space ......................... 5 cu ft
Front seats ........................................ bucket
Recliner type ...................... infinitely adjustable
General comfort .................poor fair good **excellent**
Fore-and-aft support .............poor fair good **excellent**
Lateral support ...................poor fair good **excellent**

### SUSPENSION
F:...... ind, MacPherson strut, torsion bars, anti-sway bar
R:...... ind, semi-trailing arms, torsion bars, anti-sway bar

### STEERING
Type ................................. rack-and-pinion
Turns lock-to-lock .................................. 2.9
Turning circle curb-to-curb ..................... 34.0 ft

### BRAKES
F:................................. 11.1 x 0.9-in vented disc
R:................................. 11.4 x 0.9-in vented disc
Power assist ................................... vacuum

### WHEELS AND TIRES
Wheel size ................. F: 6.0 x 16 in; R: 7.0 x 16 in
Wheel type ........................... forged aluminum
Tire make and size ..... Goodyear NCT, F: 205/55VR-16; R: 225/50VR-16

## CAR AND DRIVER TEST RESULTS

### ACCELERATION
| | Seconds |
|---|---|
| Zero to 30 mph | 1.9 |
| 40 mph | 2.9 |
| 50 mph | 4.1 |
| 60 mph | 5.3 |
| 70 mph | 7.2 |
| 80 mph | 9.0 |
| 90 mph | 10.9 |
| 100 mph | 13.9 |
| 110 mph | 17.1 |
| 120 mph | 23.1 |
| 130 mph | 29.7 |
| Top-gear passing time, 30–50 mph | 7.4 |
| 50–70 mph | 7.9 |
| Standing ¼-mile | 13.9 sec @ 100 mph |
| Top speed | 149 mph |

### BRAKING
70–0 mph @ impending lockup............... 184 ft
Modulation ...............poor fair good **excellent**
Fade ......................... **none** moderate heavy

Front-rear balance ................... poor fair **good**

### HANDLING
Roadholding, 282-ft-dia skidpad ............. 0.80 g
Understeer ............. **minimal** moderate excessive

### COAST-DOWN MEASUREMENTS
Road horsepower @ 50 mph .............. 13.0 hp
Friction and tire losses @ 50 mph ......... 5.5 hp
Aerodynamic drag @ 50 mph ............... 7.5 hp

### FUEL ECONOMY
EPA city driving ........................... **20 mpg**
EPA highway driving ...................... **32 mpg**
EPA combined driving ..................... **24 mpg**
*C/D* observed fuel economy ............... **17 mpg**

### INTERIOR SOUND LEVEL
Idle ........................................... 59 dBA
Full-throttle acceleration .................. 83 dBA
70-mph cruising ............................ 75 dBA
70-mph coasting ........................... 74 dBA

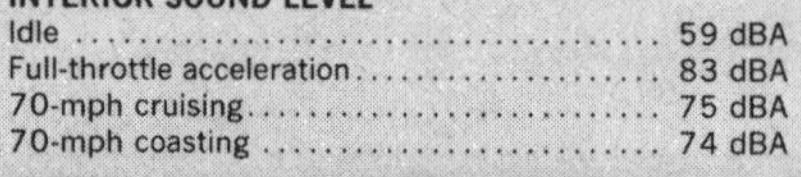


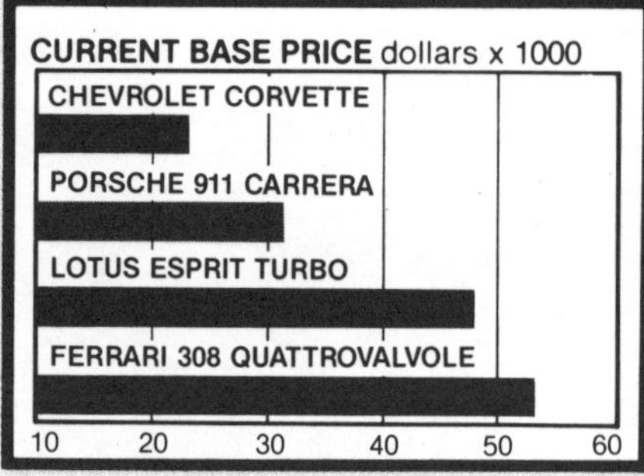


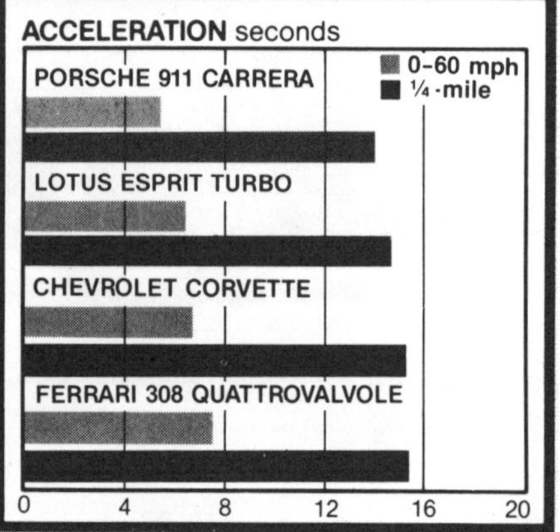


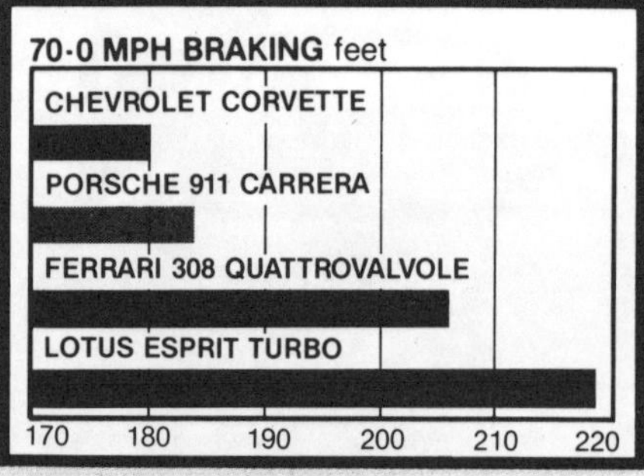


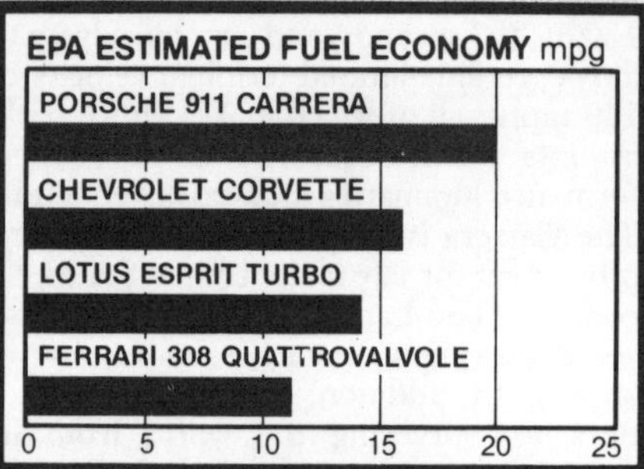


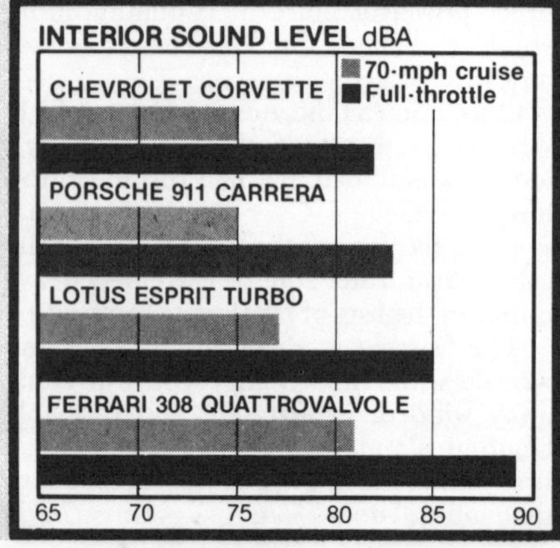

# Inside Weissach

*Where Porsche does engineering to go.*

BY PATRICK BEDARD

● Except for a touch of aristocracy in its pricing, Porsche is very much a populist car company. It believes in being close to the folks. As the huge Frankfurt international auto show kicked off last September, Porsche was the only automaker to leave its glamour cars right out on the floor—not even a rope around them—where any kid with a lollipop could come right up and do his darnedest. The pearlescent Gruppe B prototype, never before seen by non-factory eyes, was there with its doors invitingly open. A similarly pearly 956 racer upholstered in genuine leather was just spitting distance away. And so were the Porsche heavies: Peter Schutz, the managing director, Helmuth Bott, chief of research and development, and even Dr. Ferry Porsche himself. You could walk right up and have at these guys. That's what they were there for. Being available is sort of an unwritten company policy.

But when you tell them you want to visit Weissach, that's another story. Eyes roll upward. "That would be difficult," they say, "very difficult."

NASA has Houston, our atomic scientists have Los Alamos, and Porsche has Weissach (say, "Vy-sock"). Engineering is there, as are the experimental shops, the test track, and the racing department. Everything new at Porsche comes from Weissach. An enthusiast would make the pilgrimage just to warm himself in the technical heat. And if he happened to glimpse an ill-concealed secret project or two, well, heh, heh, that's what these little forays are all about.

The Weissach on everybody's map is a postcard-peaceful village about 25 kilometers from Porsche's headquarters in the Stuttgart suburb of Zuffenhausen. The roads along the way are narrow, and they loop randomly over the wrinkled quilt of Swabian farm country, cresting occasionally to reveal a tractor or some more sinister implement in immediate need of dodging. "Porsche roads," quips our host, as we reel in the distance in an arrest-me-red 944. At one crest, we glimpse in the middle distance a complex of industrial buildings. Draped over the ridge in the buildings' back yard is an area that's carefully fenced to keep the neighboring agriculture out. That's about all the fence can accomplish, because the lay of the ridge is such that, from any approach, you can see tortuous blacktop roads, fresh excavations, various shuttling cars, and numerous construction projects. Porsche's Weissach may be hard to penetrate, but for an outsider to keep it under surveillance requires little more than staying awake.

This is a relatively new property, carved out of virgin farm land in 1970, soon after Porsche was freed of its exclusive consulting contract with Volkswagen. The decision had been made to use this new liberty to become a free-lance R&D concern for hire to all comers, automotive or otherwise, and Weissach was to be the engineering center.

A guard lifts a barrier, allowing us to drive in. Porsches are everywhere. You've never seen so many. Front-engined Porsches, rear-engined Porsches, red ones, white ones, maroon ones, even one with no paint at all, left bare as a test of the galvanized-steel body (it looks awful, but not rusty). There are rows of parked Porsches and more idling around looking for a place to park. Sixteen hundred employees pour into this place every day, and after a while, you begin to realize these jellybean-colored sports cars, so highly prized elsewhere, are *transportation* here. The staffers would take a 930 off for hamburgers if the company canteen weren't the best dining room around.

Architecturally, the theme of Weissach seems to be expedience. When the place was still in the planning stages, the employees were given the choice of which they wanted first: permanent offices, or tracks and experimental shops. It was no contest. So the desk work was done in wood-framed

### Weissach Record Book

Weissach has several racetracks for both street- and race-car development. Below are the lap records for Porsche's current production cars and several important Porsche race cars of the past two decades on the fastest Weissach track, the 1.57-mile-long "Can-Am" circuit.

| Production cars | Lap speed, mph |
| --- | --- |
| 924 | 78.2 |
| 944 | 81.0 |
| 911SC | 84.4 |
| 928S | 82.6 |
| 930 | 85.7 |
| **Racing cars (driver)** | |
| 917L (Elford) | 109.3 |
| 917K (Kauhsen) | 110.0 |
| 903/03 (Elford) | 109.3 |
| 917/10 (Donohue) | 120.0 |
| 936/76 (Braun) | 115.8 |
| 935/77 (Schurti) | 111.5 |
| 956/83 (Ickx) | 123.9 |
| McLaren-TAG F1 (Lauda) | 123.6 |

barracks while the engineering facilities were put into place. Leaving the main office building for last also left time for it to be scienced to the rafters. It's a modular sort of thing, a series of glass-sided hexagons standing one against another—if more space is needed, throw up another hexagon (they were, in fact, adding one during our visit). These aren't ordinary glass-sided hexagons, however. They have bright-orange Venetian blinds hanging on the *outside*. And the blinds are wired to an electronic brain somewhere that raises, lowers, and tilts them according to its own

view of what's right for the inmates. You can imagine what sports-car engineers with a strong preference for manual steering and U-shift-it transmissions think about such automation. Even visitors can hear the murmuring.

Weissach is your basic full-service proto-

*Peter "Columbo" Falk, Porsche's head of racing, points out the most photogenic curve on the Can-Am course, where we later catch a 956 under test.*

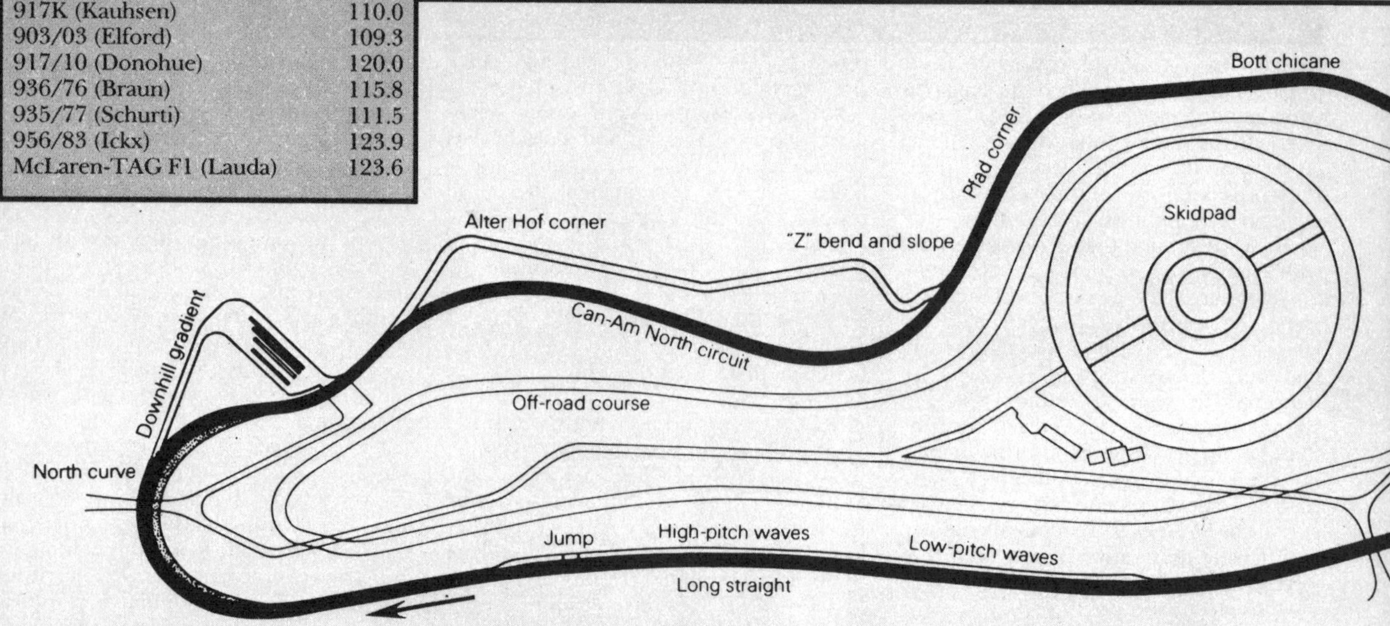

type shop. The staff can design pretty much anything you might want, and they can also make it. The premises even include a foundry to pour both aluminum and iron. Brake discs and gearboxes for the racing cars are done here. In fact, the 956 series of racers, which has won Le Mans for the last two consecutive years, is built and serviced at Weissach.

Before we're taken on a walk through the experimental garage, we're encouraged not to look too closely at the pieces lying around and to forget entirely the non-Porsche vehicles we might see. Not, mind you, that seeing some other make necessarily means anything; the engineers might merely be tearing apart something that catches their fancy, just to see what makes it work. But there is no hiding the fact that a lot of development is going on around here for other companies—that's a business that Porsche pursues—and confidentiality is part of the deal. This engineering-to-go concession isn't a huge moneymaker, maybe only fifteen percent of annual sales, but it's what allows Porsche to keep pushing the outside of the technical envelope

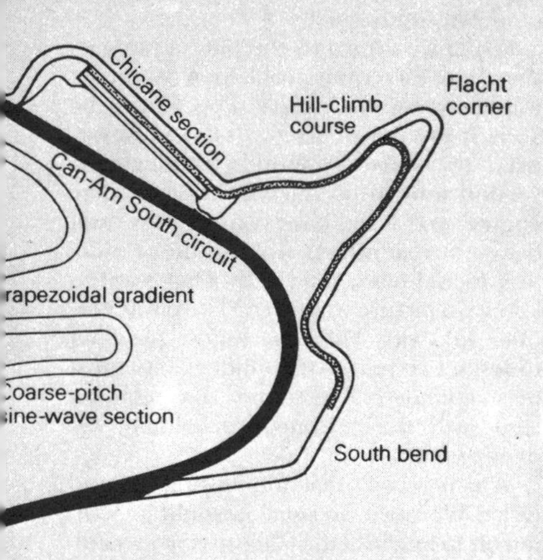



with its own street and racing cars. Projects like the Group B require real engineering horsepower when they're in the development stage, but such projects also come and go. By selling off its surplus engineering time, Porsche can keep a far larger staff than any other low-volume builder. This consulting business is one of the reasons Porsche confines its model lineup to sports cars: they do not compete with the products of its engineering customers. Porsche doesn't talk about its customers, either, because most of them want the world to think they come up with their own ideas. But some like the prestige: Yamaha has advertised its Porsche connection. The Russians talk openly about Porsche's contribution to the new Lada. It's also known that Harley-Davidson had an engine done here and the West German Leopard tank spent a lot of time at Weissach. Nonetheless, we're sworn to secrecy.

Inside the shop door, about the second thing we encounter is the airplane engine based on the 911 six-cylinder. This is deemed public knowledge, so we can mention it. There are engines, gearboxes, and odd parts everywhere, rather like a meat market specializing in aluminum castings. And, of course, cars—Porsches and others—nosed into service bays and up on hoists. Racing is never far below the surface at Weissach. Stacks of mounted tires are lined up at the wheel-balance department. About a third of them are fat Dunlops on corroded magnesium center-lock wheels. There is only one job for such hardware.

The racing department is next. It's well away from the main buildings, so close to the skidpad the commute would barely get the oil pressure up. The office building is wood painted dull red. In the corridor is a poster advertising a forthcoming tennis tournament; Martina Navratilova, in a Porsche T-shirt, is stretching for a ball. If you're worried about the Porsche factory racers being technocratic monks, forget it: she's sporting chin whiskers, and nipples have been penciled onto her shirt. We continue down the corridor past the Dinkel Acker vending machine (deposit DM 1.10) and turn left, entering a room chaotic enough to be an editorial office except there are too many drilled brake rotors slid under desks and too many wind-tunnel models parked atop the file cabinets. Peter Falk will talk to us. He's the head of Porsche racing, a grandfatherly man as thin as asparagus, and he obviously didn't take the job just for the double-wide broom closet he was given for a private office. He tells us that about 60 percent of the race-car testing is done right here at Weissach. They'd do more here except that the track doesn't have enough straightaway for high speeds. The 956 runs out of room at 280 kph (174 mph), well short of the 370 kph (230 mph) it reaches at Le Mans. When they need to really let a racer out, they go to Paul Ricard in France. The concentric skidpads, however, are very useful for both low- and high-speed testing, particularly the 210-meter (689 feet) outer circle, which allows enough speed for aerodynamic tuning. Newly designed racers are also tested over the torture roads, which is probably why Porsches have so few structural failures on the track.

The resources that didn't go into Falk's

# The Secret Sharers

• Although Weissach exists primarily for the development of Porsche's production cars and for its motor-racing activities, its use by outside customers is on the increase. In the fiscal year that ended last July 31, annual billings for such use rose from $23.5 million to more than $31 million.

"Over 80 percent of the company's 40 to 45 clients are directly or indirectly related to the motor industry," says Horst Marchart, the managing director in charge of all outside R&D. "Manufacturers are generally happy with what we do for them, but many don't like talking about the helping hand in the background." Marchart adds, "To be quite honest, there are one or two clients *we* would not like the Porsche name to be associated with, either."

The engineers at Weissach aren't naming names, but we have managed to learn a few of them nonetheless. The current customer list includes:

—**Aerospatiale:** A310 Airbus cockpit design.
—**Audi:** bench and durability testing, engine development.
—**Ford:** engines, cylinder switch-off mechanism, chassis development and testing.
—**West German Ministry of Defense:** Leopard tank and Biber and Wiesel special-purpose vehicles.
—**Harley-Davidson:** engine.
—**Lada:** development of a compact fwd car from drawing board to production.
—**Lycoming:** aircraft-engine development.
—**Peugeot/Talbot:** gas and diesel engines, turbocharging, chassis work.
—**SEAT:** development of a small fwd hatchback, component design.
—**Volkswagen:** bench and durability testing, component development.
—**Yamaha:** motorcycle-engine development.

*—Fred Dillinger*

*Porsche's Group C 956 grabbed nine of the top ten finishing spots at Le Mans last year and may compete in IMSA this year. The factory fleet comes home to Weissach for refurbishing.*

office were put into the race shop itself. Huge overhead doors open to three Rothmans-liveried 956s stripped to their skeletons for rebuild in preparation for their next appearance, at Kyalami. There is a quiet confidence in the air, and little attention is paid to us. This building is surely Porsche racing's center of gravity. A ten-long row of sun-faded victory wreaths hangs on the wall over a flank of workbenches. Through a doorway is the white-walled engine room. It's brightly illuminated, with rows of green-topped benches separating the 956 rebuilding area from the TAG Formula 1 engine area from several other areas that we can't identify. Never having seen any of these engines apart before, we're poorly prepared to spot secrets. Falk probably factored that in when he okayed the tour.

We've been waiting for the skidpad to dry up so we can get on with photographing the Group B prototype, which has been temporarily delayed from its wind-tunnel appointment just for our purposes. But it's a gray, cold day—back in Zuffenhausen, Weissach is called "Siberia," not for its remoteness, but because the temperature is typically five degrees Celsius colder here—and the puddles in the wavy asphalt simply won't go away. Our ever resourceful photographer proposes making them bigger and using them as reflecting pools. So we need more water. Off to the side of the skidpad is a low wooden shed with grounds-maintenance equipment parked around it. Surely there will be water there. On the way over, Manfred Jantke, chief of the press office, mentions that just this side of the building is the spot where all the stationary testing of the aero engine was done. A propeller was attached, and it just roared away right there in the open air. But as we draw near, he's surprised to find the location is bare. Not even an oil spot. That, Jantke says, is typical of the improvisational nature of Weissach. The engineers just lash up a rig that will get them the answers. When the job is done, there are no monuments. It's the same temporary theme so evident in Falk's barracks office.

As we near the shed, a woman in her late twenties steps out. She's wearing a red Porsche jacket. Jantke asks in German about the water. I poke my head into the doorway, expecting to see shovels and bags of concrete. Surprise. It's a makeshift laboratory. There are banks of chart recorders in the foreground sprouting electrical cables trailing back to what look like highly instrumented dynos. A motorcycle engine is mounted onto one, and a 928 aluminum V-8 onto the other. On closer inspection, the V-8 is actually a test fixture, its valley cut away to reveal combustion chambers with four valves, twice what you'd find in the production version. The water, we are told, is outside.

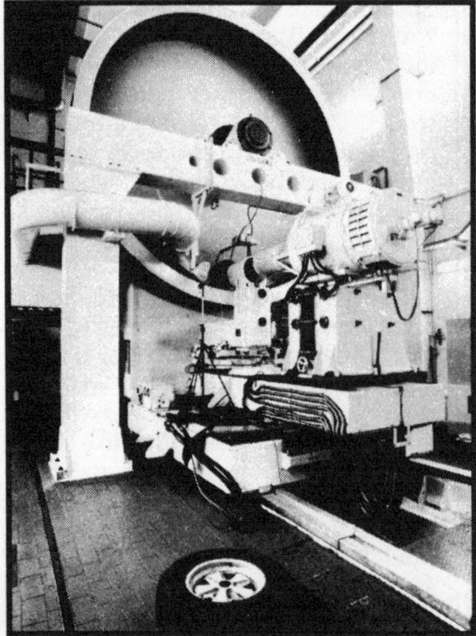

*Porsche is one of the few firms in the world with its own tire-test machine. The twenty-foot drum can measure wet and dry traction.*

The guided tour includes a few laps around the "Can-Am" circuit, where Mark Donohue's 917/30 practiced its moves a dozen years ago before stepping into the open and devastating the opposition in the old Can-Am. It's 1.6 miles around with six turns, including a steeply banked bowl at the south end and a flatter U-bend at the north, which is separated from the main engineering compound by only a few rows of catch fencing and a weedy dirt bank. Miss this one, which comes right after the high-speed straight, and you'll end up in the emissions-durability test section, probably dead. Once we get going, it's obvious why racing drivers dislike testing here: there's zero runoff room. At Alter Hof, which is technically not part of the Can-Am circuit, and at Pfad corner and Bott chicane, which are, the track runs right along the edge of the property. The catch fencing starts where the pavement stops. And the pavement is narrow to begin with. As we exit full-throttle out of Pfad, our driver inclines his head toward a spot in the fence and says that a 956 glanced off there a few weeks ago and comprehensively crunched itself against a rail on the inside of the track. What about the pilot? Oh, he's that guy we saw around the race office with the cast on his wrist. The wheel spun around and wrenched his thumb.

We alternate between a 928 and a 930. With European-spec horsepower, both of them fly, up the slope (it seems as steep as a stairway) at "Z" bend, sliding the tail wide for the photographer, then through the narrows to the banking at the south. On the long straight, as these street cars reach well up in top gear, there's time to check out the orchard that butts against the track on the west. The terrain falls away so that we are about mid-height on the trees. Then we brake hard to stay out of the engineering compound, arc under the bridge connecting the office area to the track, and power on down south again.

When we return to the photographer's site, there's a serious problem. A prototype wandered by that wasn't supposed to be seen. It was not a Porsche. It did, however, make delicious V-8 sounds, although the version now in production has an in-line engine; and it did have conspicuous camber in the rear wheels, whereas the production model has a solid axle. They want to know if a picture was taken. The photographer says no. Then, in almost pleading tones, we're told that confidentiality must be maintained. The engineering relationship with the customers absolutely depends on it.

Which means that this testing ground called Weissach, so small it would be written off to infield at the GM proving ground, will be largely left anonymous. It has no shiny tools and towers inside to stand as monuments to Porsche engineering, and all but a few of the monuments on the outside go about wearing other nameplates. ●

# Porsche 911 Group B

*The good doctor's dream on its way to immortality.*

BY DON SHERMAN

• You can tell the car manufacturers with vision—they're the ones who never look back. Porsche is a prime case in point. Last fall's Frankfurt Auto Show marked the 911's twentieth birthday, and to celebrate the occasion, the wizards of Weissach unveiled a new Group B design study with enough pure technology to sustain the rear-engined throwback indefinitely.

Porsche's managing director, Peter Schutz, is doing everything in his power to grant the 911 immortality, because he knows full well it is the center of Porsche enthusiasm. To keep those faithful to the 911 appeased, Schutz has assured them that their car could stay in production another ten years. (Dr. Porsche asks, "Why not twenty?") This means that the technological advances previewed at Frankfurt—four-wheel drive, a twin-turbo engine, and a six-speed transmission—will gradually filter onto the assembly line to keep the 911 vital, as other innovations have for the past twenty years.

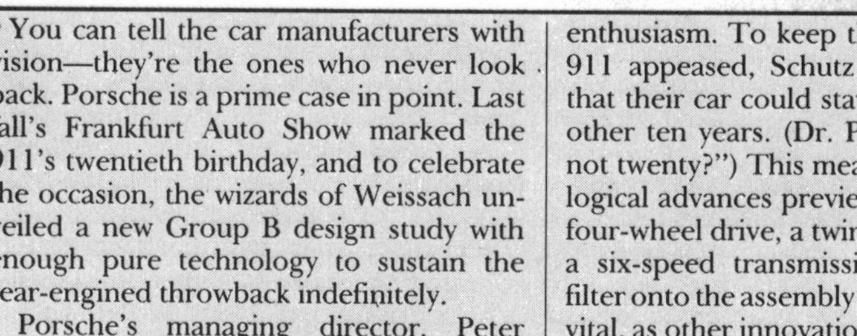

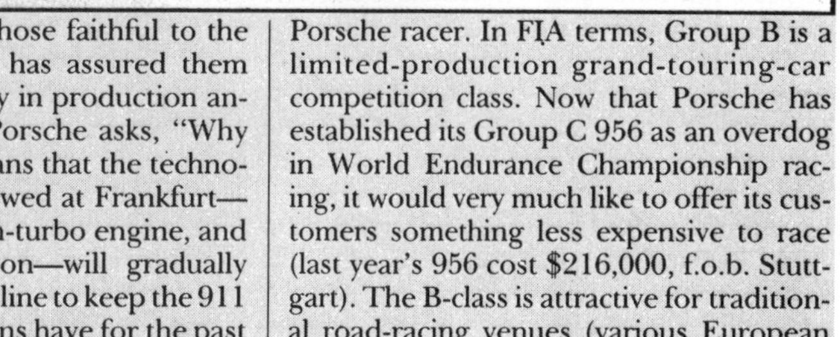

The pearlescent soap bar previewed at Frankfurt is also the prototype of a future Porsche racer. In FIA terms, Group B is a limited-production grand-touring-car competition class. Now that Porsche has established its Group C 956 as an overdog in World Endurance Championship racing, it would very much like to offer its customers something less expensive to race (last year's 956 cost $216,000, f.o.b. Stuttgart). The B-class is attractive for traditional road-racing venues (various European championships, IMSA GT, SCCA Trans-Am) and also for rallying. No-holds-barred battles among the Audi and Lancia fac-

tories and various semi-independents over the past few years have piqued rally competition to a fever.

In both spectator interest and intensity of competition, rallying is outranked only by Formula 1, and Porsche would love to give its customers an opportunity to enter the fray. The factory is quick to point out that it has no intention of jumping feetfirst into world-class rallying with a cost-is-no-object in-house effort. Instead, Porsche plans to build the machinery for and sell it to independent customers. The competition is so strong and the logistics of far-flung rally events are so complicated that an enormous budget will be required for any chance of success.

Showing the world state-of-the-art technology and building highly competitive race cars is strictly business-as-usual for Porsche. The Group B study does, however, mark one significant kink in its path: an undeniable commitment to four-wheel drive. More than two years ago, at the 1981 Frankfurt show, similar handwriting was on the wall in the form of a four-wheel-drive 911 Cabriolet show star. Shortly thereafter, the Cabriolet did go into production (with a conventional powertrain), while Porsche offered only vague plans for a future four-wheel-drive model. Now that the Group B prototype has seen the light of day, however, a good deal more is known. The factory is finally frank about its interest in and commitment to four-wheel drive, so we wasted no time jetting a *C/D* team to Stuttgart-Zuffenhausen to investigate.

Manfred Bantle, the Group B project manager at Porsche's Weissach development center, admitted that, while the design of the Frankfurt display began less than a year ago, the *Werke* has had an intense interest in four-wheel drive for years. A Jensen FF was purchased in the mid-Sixties to allow a detailed analysis of the Ferguson-patented four-wheel-drive system. Herr Bantle remembers the project well because it was one of his first experiments at Porsche. With typical German thoroughness, the Jensen was modified to allow testing with front drive, rear drive, and four-wheel drive. Skidpad studies revealed that rear drive was the fastest system for steady-state cornering, four-wheel drive was second best, and front drive was the slowest. In high-speed tests with drivers of various skill levels, Porsche engineers found that a good driver typically did best with rear drive, while poor drivers were more comfortable and capable with four-wheel drive. More recently, Weissach teams have spent their winters evaluating prototype drivetrain hardware over ice, snow, and gravel conditions in the Austrian Alps. The Falkertsee frozen lake and the Türracher Höhe mountain pass are two favorite locations, and Porsche's chief of research and development, Helmuth Bott, is said to be a most enthusiastic 4wd test pilot.

Porsche engineers have also kept close tabs on world-championship rallying through their ties with Audi. Virtually every prospective entrant now considers four-wheel drive an absolute must for success, even though Lancia did seize the laurels last year with its rear-drive Rally. (It will reappear with 4wd this year.) The pitched battles between the mid-engined Lancia Rallys and Audi's front-engined, four-wheel-drive Quattros gave Porsche engineers a wonderful opportunity to test some of their pet theories from the sidelines. Bantle says it's well known that a rear-driver's nose can be twitched toward the center of a curve just by breaking the tail loose with power. The four-wheel-drive machines shift from understeer to oversteer in a much smoother fashion, so adding power has a completely different effect on the car's attitude. Step on the throttle with four-wheel drive, and the whole car drifts wide. Since the nose won't point in, drivers instead aim the front wheels in the desired direction of travel with the steering wheel. The churning front tires help pull the rest of the car into the bend in a dog-follows-the-leash fashion.

Porsche believes a superior approach would be to avoid the understeer altogether. The nose-heavy Quattros sometimes plow like farm implements. Mid-engined Lancia Rallys are superior in this regard, but the rear-engined, four-wheel-drive 911 should offer a significant advantage over both competitors. Alpine tests have been most encouraging in this respect, and the Porsche factory is quite anxious to charge ahead with phase two, a competition debut. Jacky Ickx has asked the factory to build a fleet of three normally aspirated, four-wheel-drive 911s for this year's Paris-to-Dakar Rally, and the proposal has struck a responsive chord at Weissach. Ickx will drum up the sponsorship funds and the

necessary support army, while Porsche supplies the technology. Ickx will pilot one 911, a pair of Frenchmen will drive and navigate the second one, and two hard-charging mechanics will man the third entry. The strategy is to keep the mechanics close at hand to help Ickx if he falters en route out of the reach of the normal service crews. Likewise, the Frenchmen's mount will be cannibalized if necessary to improve Ickx's chances for success. In the event of a win or a strong finish, Porsche will of course share the glory, but it will also benefit from the grueling durability test. The experience will surely help shape the Group B rally machine planned for the 1985 season.

The other half of the Group B's future competitive life will be road racing. Here, the path ahead is not as well mapped, because everyone in the past who has tried to campaign an all-wheel-drive road racer has failed miserably. Nevertheless, Porsche is anticipating the monumental challenges ahead with much the same enthusiasm as an Einstein sitting down to unravel relativity theories. The Weissach team feels the Group B machine will be an ideal way to probe the misty unknowns of hard-pavement traction and handling.

Theoretical studies suggest that all-wheel drive will be a great aid to accelerating out of turns with ultra-high-output engines. (The pearly-white show car's tail was packed with a 400-horsepower 956 engine in hydraulic-lifter "street" trim. A competition version would carry much more power.) In addition, Herr Bantle expects that the chassis's response to the throttle will be much smoother than is possible with a rear-driver, and there will be no abrupt change in cornering attitudes at the limit of adhesion. He isn't depending on four-wheel drive to deliver any gains in lateral grip, as that has historically and will in the future come from aerodynamic downforce. The Frankfurt show car has already made a number of trips to the wind tunnel, and an 0.32 drag coefficient has been established as a developmental goal. The adjustable rear wing will help balance the aerodynamic downforce produced by the Group B's smooth shape, while ground-effect bodywork is also a possibility.

One critical component that hasn't progressed much beyond the blue-sky-idea phase, as yet, will eventually be buried deep within the Group B's powertrain. Porsche plans to use an electronically controlled center differential to split the torque between the front and rear wheels. It will react dynamically to control balance, depending on track conditions and driver inputs. Conceivably, there could be a plug-in chip that optimizes a host of variables for, say, Daytona, and a drastically different drivetrain program for the Nürburgring. A fine-tuning knob might also be provided so the driver could alter the power delivery in much the same way he can today shift braking effort or roll stiffness in a contemporary race car.

The Group B car is a major departure from past roadgoing 911s not only in its exotic powertrain but also in its chassis design. The huge brake rotors and calipers are right off the 956, there's not a Mac-Pherson strut or a semi-trailing arm in sight, and there are shock absorbers everywhere. Wheel loads feed into the unit-construction body through unequal-length control arms and two dampers per corner. Each front shock has a concentric coil spring, while the rear axle has only one spring per wheel. The rack-and-pinion steering and the anti-sway-bar equipment are in no way remarkable, but there is an interesting toe-control link between each rear suspension hub and its lower locating arm. Herr Bantle was mum on the subject, but the links' steep angle of attachment appears to be a means of factoring in some rear toe-in during hard braking and hard cornering to improve stability. (The 928's Weissach axle accomplishes similar ends.) For more detailed explanations, we may have to wait until the next Frankfurt show.

We'll also have to be patient while Porsche goes through the painstaking task of developing the Group B show car into a working machine. The Jacky Ickx program will doubtless offer some preview of the B's performance potential, but that's all we're likely to see through 1984. As you can imagine, Porsche engineers do their best work in the privacy of their own test sessions. If the hardware is ready near the end of the year, an elite assembly crew will be formed at Weissach to bolt together 200 Group B 911s for FIA homologation. The most potent ones will be sold to Porsche's racing and rallying customers with the right stuff (equal measures of money and competition success), while a few "street" versions will be offered to well-heeled civilians. We understand that several of the most resourceful Porsche dealers are already accepting deposits.

**Vehicle type:** rear-engine, 4-wheel-drive, 2-passenger, 2-door prototype
Engine type: twin-turbocharged and intercooled flat 6, aluminum block and heads, Bosch Motronic engine-control system
Displacement . . . . . . . . . . . . . . . . . . . . . 174 cu in, 2850cc
Power (street trim) . . . . . . . . . . . . . . . . . . . . . . 400 bhp
Transmission . . . . 6-speed with electronic center differential
Wheelbase . . . . . . . . . . . . . . . . . . . . . . . . . . . . . . 89.4 in
Tires . . . . . . . . Dunlop SP Sport Denloc, F: 235/40VR-17; R: 255/40VR-17
**Factory performance ratings:**
Aerodynamic-drag coefficient . . . . . . . . . . . . . . . . . . . . 0.32
Zero to 62 mph . . . . . . . . . . . . . . . . . . . . . . . . . . . . 4.9 sec
Zero to 124 mph . . . . . . . . . . . . . . . . . . . . . . . . . . 15.4 sec
Top speed . . . . . . . . . . . . . . . . . . . . . . . . . . . . . . 186 mph

# The Best-Handling Imported Car Is…

*Yet another Angeles Crest fest.*

● And you thought we had risked it all racing up and down treacherous mountain passes to name the Camaro Z28 the best-handling American car (*C/D*, May). Well, we're back on the rock, dancing with the same seraphim of the Angeles Forest, but we've upped the ante to *eight* contestants—over a quarter of a million dollars' worth of turbocharged, four-valved, double-overhead-cammed heart stoppers. The part that hasn't changed much is the mission. We've brought imported machines to the mount this time, but the challenge once again is to find the best-handling automobile of the lot.

As before, this is a pure roadability investigation. No acceleration figures or braking distances were allowed to confuse the proceedings. Price was not a factor, either: a quick check of the vital statistics will reveal window stickers ranging from $10,345 to $50,000. Any current model that enters the United States on a passport was a potential contender.

At the start we had no intention of testing eight cars to ascertain the best import, but once the floor was open to nominations, the kid-in-a-candy-store phenomenon took over. We wanted representatives from all over the globe, a mix of the affordable and the exotic, and a variety of powertrain layouts. We purposely sought out several highly reputed brands and a couple of Johnny-come-lately challengers. Ferrari, Lotus, and Porsche *had* to be represented if this was to be a legitimate exercise, but that raised the question of *which* Porsche. We answered it by inviting all three of Ferry's finest. The Audi Quattro is one of the boldest handling experiments ever to press rubber to the road, so we had to have the four-wheel-driver, even though only a handful are sold here each year. Since front-wheel drive made such a great impression in our last test, it seemed imperative to consider a front-driver this time around. Before we knew it, our best-handling shopping cart contained a two-door sedan and seven coupes; an all-wheel-driver, a front-driver, and six rear-drivers. Front-, mid-, and rear-engined layouts are all represented, though we did not make room for a mid-engined, front-drive station wagon. Maybe next time.

Rounding up the rolling stock and the personnel to keep it rolling was no mean feat in itself. Ferrari North America bowed out, claiming that its entire West Coast stock of Quattrovalvoles had been sold! Fortunately, Mike Sheehan of European Auto Restoration in Costa Mesa, California, stepped into the breach to lend us a 1984 coupe off his gray-market lot. A Quattro had to be trucked in from a Gulf Coast port at the last minute, so of course the driver fell ill on the road. And although

LOTUS ESPRIT TURBO

FERRARI QUATTROVALVOLE

we intended to keep the original six-person jury intact for this second round of tests, Jean Lindamood missed the action because of an injury. (Her separated shoulder is on the mend.) The sixth seat was capably filled by Patrick Bedard, *C/D* editor at large.

The battery of subjective and objective handling tests tooled up for the American-made cars had worked so well that little fine-tuning was deemed necessary for this round. We allotted more time to study on-center handling, and we applied a new photoelectric timing device to several of our track tests. Since the SCCA gymkhana course at the Chrysler-Shelby Performance Center had become a tractor-trailer parking lot, we laid out our own array of cones at the Los Angeles County Fairgrounds.

As usual, you'll find a blizzard of performance statistics sprinkled throughout the next few pages, but don't pay *too* much attention to individual test scores or race-track results. What's most important here is the bottom line; we've determined the best-handling import with a single, easy-to-understand *subjective* vote. The path to that determination starts right here.

### The Test Track

Confirming impressions is what the test track is all about. It's a safe place to explore the extremes of a car's behavior. In the process, we define performance limits and probe handling characteristics that can't be fully studied on the street.

We put the best-handling-import contenders through the same track tests to which we subjected the domestic candidates in May, at the same test facility. These included both smooth and rough skidpads, a 900-foot slalom, a lane-change maneuver, combined acceleration and cornering, and combined braking and cornering. We made a few minor procedural alterations, but the results reported here should be comparable with the measurements taken two months ago. We also ran each car through an SCCA Pro Solo gymkhana course, as before, but a slightly slower lay-out than the one we used in May.

The Porsche 911 Carrera with the optional Turbo chassis took top honors in both skidpad tests, managing 0.84 g on the smooth circle and 0.82 on the bumpy one. In each case, it displayed the classic 911 traits of power-on understeer and power-off oversteer, with a fine line between the two. The Carrera's ability to absorb the rough circle's peaks and valleys without ever bottoming its suspension was most impressive.

The Porsche 944 displayed similar characteristics, but with a softer transition between oversteer and understeer. It circulated at 0.82 g on the smooth pad and at 0.81

| | price, base/as tested | powertrain | engine/ transmission | suspension | |
|---|---|---|---|---|---|
| | | | | front | rear |
| **AUDI QUATTRO** | $35,000/$35,575 | front-engine, 4-wheel-drive | turbo 2.1-liter 5-in-line/ 5-sp manual | ind, MacPherson strut, coil springs, anti-sway bar | ind, MacPherson strut, coil springs, anti-sway bar |
| **FERRARI QUATTROVALVOLE** | $50,000/$50,000 | mid-engine, rear-drive | 2.9-liter V-8/ 5-sp manual | ind, unequal-length control arms, coil springs, anti-sway bar | ind, unequal-length control arms, coil springs, anti-sway bar |
| **HONDA PRELUDE** | $9995/$10,345 | front-engine, front-drive | 1.8-liter 4-in-line/ 5-sp manual | ind, unequal-length control arms, coil springs, anti-sway bar | ind, MacPherson strut, coil springs, anti-sway bar |
| **LOTUS ESPRIT TURBO** | $47,984/$47,984 | mid-engine, rear-drive | turbo 2.2-liter 4-in-line/ 5-sp manual | ind, unequal-length control arms, coil springs, anti-sway bar | ind, 1 trailing arm and 2 transverse links per side, coil springs |
| **PORSCHE 911 CARRERA** | $31,950/$45,880 | rear-engine, rear-drive | 3.2-liter flat 6/ 5-sp manual | ind, MacPherson strut, torsion bars, anti-sway bar | ind, semi-trailing arm, torsion bars, anti-sway bar |
| **PORSCHE 928S** | $44,000/$45,820 | front-engine, rear-drive | 4.7-liter V-8/ 4-sp auto | ind, unequal-length control arms, coil springs, anti-sway bar | ind, unequal-length control arms, coil springs, anti-sway bar |
| **PORSCHE 944** | $21,440/$23,145 | front-engine, rear-drive | 2.5-liter 4-in-line/ 5-sp manual | ind, MacPherson strut, coil springs, anti-sway bar | ind, semi-trailing arm, torsion bars, anti-sway bar |
| **TOYOTA CELICA SUPRA** | $15,724/$16,809 | front-engine, rear-drive | 2.8-liter 6-in-line/ 5-sp manual | ind, MacPherson strut, coil springs, anti-sway bar | ind, semi-trailing arm, coil springs, anti-sway bar |

*Vital Statistics*

PORSCHE 911 CARRERA

PORSCHE 928S

g on the rough circle, but in the latter test it suffered much bobbing and an occasional bottoming crash. The Toyota Supra and the Porsche 928 behaved in a similar fashion, though both demonstrated less grip. The Ferrari scored in the same group, but once its tail swung beyond a certain point, there was no catching it; its loose rear end was especially noticeable on the bumpy circle, where its slow, heavy steering wasn't much help in managing its wayward tail. The Lotus Esprit Turbo also tended to swing its tail in lift-throttle situations; its steering would go light at the same time, confusing the driver by implying an understeer that wasn't really present.

In contrast to these cars, the Honda Prelude and the Audi Quattro cornered purely in an understeer mode. Both could be pushed to their limits under power and

| curb weight, lbs | weight distribution, % F/R | wheelbase, in | tires |
|---|---|---|---|
| 3060 | 58.2/41.8 | 99.5 | Pirelli P7R, 215/50VR-15 |
| 3000 | 41.3/58.7 | 92.1 | Michelin TRX, 220/50VR-390 |
| 2300 | 61.7/38.3 | 97.7 | Bridgestone RD-116 Steel, 185/70SR-13 |
| 2700 | 42.2/57.8 | 96.0 | Goodyear NCT, F: 195/60VR-15; R: 235/60VR-15 |
| 2850 | 39.3/60.7 | 89.4 | Pirelli P7, F: 205/55VR-16; R: 225/50VR-16 |
| 3440 | 50.6/49.4 | 98.4 | Goodyear NCT, 225/50VR-16 |
| 2800 | 47.9/52.1 | 94.5 | Pirelli P6, 215/60VR-15 |
| 3040 | 53.9/46.1 | 103.0 | Bridgestone Potenza, 225/60HR-14 |

held there by lifting off when necessary. The Prelude responded instantly to this throttle change by tucking in its nose. The Quattro behaved similarly, though more sluggishly.

Our lane-change and slalom tests both demand controllability in addition to adhesion, though the lane change is the more forgiving of the two. The Esprit provided the best way to compare the relative severity of the two tests: it was by far the fastest lane changer (thanks to its extremely quick turn-in response), but its non-linear-steering and tail-wagging tendencies hurt it in the more repetitive slalom test. The Porsche 944 exhibited the opposite characteristics: its superb controllability helped produce the fastest slalom-course speed, but it fell to fourth place in the lane-change event.

The 911 Carrera was quite capable through both courses, but only when driven below its limits. A little too much speed, and it started sliding past the point where recovery was possible. The 308 also had to be driven very carefully to its second-best slalom clocking, but its lane-change performance was hampered by slow steering. The Honda Prelude was much easier to drive in these tests, its performance being limited largely by a lack of grip.

The Quattro and the 928 were the poorest finishers in the transient tests. Both cars demonstrated their bulky and heavy natures by resisting sudden directional changes, and the effect was magnified by ponderous steering. The Quattro was further hindered by a two-step response to sudden steering inputs, which made it difficult to position accurately.

In our combined cornering-and-braking test, the Esprit again demonstrated its excellent turn-in ability, and its tail-end highjinks were easy to check in this one-shot maneuver. The Ferrari, the 944, and the Carrera also felt quite capable: all three had useful tail-out tendencies. The Supra and the 928 exhibited similar behavior, but they hung out their tails at lower speeds because of their lower traction limits.

Interestingly enough, the Quattro kicked out its tail during the turn-and-brake phase of the test, but it promptly shifted into heavy understeer when a throttle correction was applied. The Prelude was also hindered by an excessive dependence on its front tires. The feeling was secure, but also very slow.

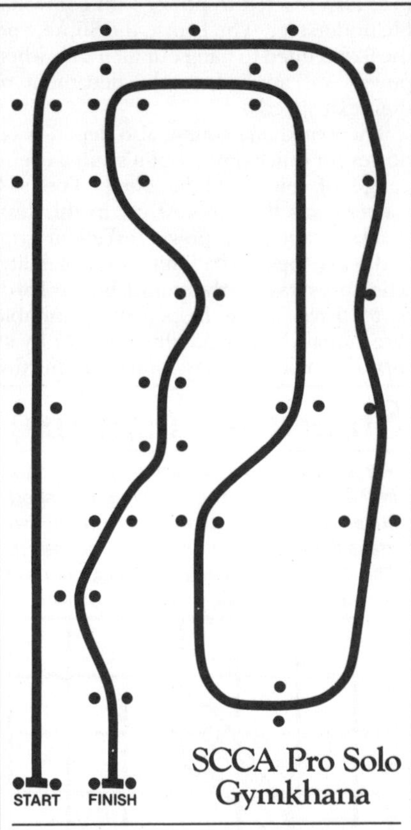


SCCA Pro Solo Gymkhana

The SCCA Pro Solo course tests low-speed handling. Quick runs require sharp throttle response, instant maneuverability, and minimal understeer, especially under braking.

TOYOTA CELICA SUPRA

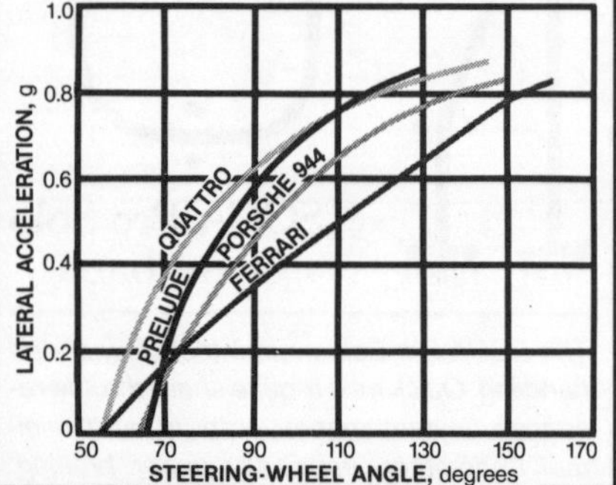

AUDI QUATTRO

The ability to accelerate out of a corner at the limit obviously depends on the power available, so it's not too surprising that the 308 was the quickest of several very quick cars in our cornering-while-accelerating test. It had enough understeer to allow an early, heavy power application. The Carrera, the Quattro, the 944, and the Prelude also reacted to power with controllable understeer. The Esprit, the Supra, and the 928 tended to hang out their tails when power was applied, to the detriment of their exit speeds.

The gymkhana course also depends on power for quick times, but it's still a useful gauge of low-speed handling. The 911 Carrera was the quickest car in this test, thanks to abundant power, excellent grip, and easy below-the-limit controllability. The Lotus was nearly as quick but far harder to drive; it was helped by its nimble directional-changing ability but hurt by its turbo lag, tail-happiness, and unsupportive

## Steering Response

*For clarity, we've shown only four cars in the plot below. The Quattro has a short linear range, while the Ferrari's response stays linear almost to its limit of adhesion. The 911, 928, Lotus, and Toyota fall near the middle between these two extremes.*



seat. The Ferrari was not far behind; it had tenacious grip and excellent power but was again hampered by its heavy, slow steering. The 944 was equally quick; it was easy to drive but suffered from too much understeer.

Surprisingly, the front-drive Prelude showed little understeer in the gymkhana. It trail-braked nicely, and its light, quick steering was perfect for the tight course. In contrast, the power-steering units in the Supra and the Quattro lost their assist in the course's tight turns, hindering their progress. The Supra also tended to spin its inside rear tire uselessly, and the Quattro didn't like the course's sudden directional changes. Nor did the 928, which alternated between excessive plowing and wagging its tail between the tight cones.

### The Racetrack

Racetracks provide a sort of theoretical test of handling. Each turn is practiced to perfection, and every inch of the road can be used without fear of oncoming traffic, which means that track operation is pretty much irrelevant to what happens on the street. But track operation also cuts through all the excuses. There is no way a car can hide its handling quirks. If it's ornery at the limit, you find out immediately.

Only a few years ago, there wasn't a production car alive that would acquit itself with honor on a racetrack; they would all fall to their knees when pushed. What's remarkable about this test is that four of the contestants behaved with true poise at the limit. (Curiously, the eight test cars fell into two groups: those with poise and those with speed. The fast ones were far more challenging to keep on the pavement, and this had nothing to do with being overpowered. They just didn't handle as well.) The high marks go to the Prelude, the Supra, the Quattro, and the 944. Willow Springs put forth its damnedest—constant-radius tire squealers, cresting turns, plummeting turns, ultra-high-speed sweepers, and the nasty, decreasing-radius Turn Nine—and these four took it all with aplomb.

The Honda was the slowest of all the cars, averaging 78.8 mph for a lap of Willow. Its modest power holds it back. Its handling, however, is delightful: very light to the touch and quick to respond. The 70-series Bridgestones aren't terribly sticky, but the car seems perfectly tailored to what grip they do have. You can use it all. The chassis is so predictable you needn't leave big margins. As you would expect of a front-driver, there is understeer with power and tuck-in when you lift. The special joy of the Honda is that it seems to have exactly the right amount of each, which is to say, not much.

The idea that Japan can't build good-handling cars is hereby put to rest. The Supra also earned a berth in the "poised" group, with a mild understeering tendency that it maintains over a broad range of operation. It's calm and stable, almost relaxing to drive hard. No bad habits wait to trip you up. Only a pedal placement that prevents heel-and-toeing keeps this car from showing its best at the track.

If this lead group proves anything, it's that good street-car handling is not the sole province of any one driveline configuration: we had a front-driver, two rear-drivers, and a four-wheel-driver. The Quattro is too big a car with too little horsepower to produce fantastic lap times, but its R-coded Pirelli P7s did supply amazing grip and its four-wheel drive ensured unflappable stability. This car has a clear and predictable preference for understeer; no matter how you hack at the wheel or play the pedals, that's what it does. If you're into a turn too hot, easing off the power reduces speed, bringing the car back into line. But there's no tuck-in when you lift, just understeer. When it comes to spin-out resistance, the Quattro easily leads the league.

There is a strong family resemblance in the behavior of the two front-engined

Porsches, yet we much prefer the 944. There probably isn't another volume-produced car anywhere that takes to the track as well. The controls are positioned exactly right, and the car is wonderfully tolerant at the limit. Understeer is gentle under power, replaced by a nice tuck-in when you lift. In part because of the gradual nature of the Pirelli P6s, the 944 is terrifically slidable. Both ends seem to let go simultaneously and then reattach themselves to the pavement in the same controllable way.

The Pirelli P7s on the 928 have a sharper breakaway; combined with the car's clear preference for tail-out cornering, they make for far trickier handling. Still, the 928 was about a second a lap faster than the 944, and only two seconds slower than the 911, the Lotus, and the Ferrari, which turned in virtually identical times. Had we been able to get a 928 with a five-speed manual transmission instead of a four-speed automatic, the car would probably

have finished closer to the top group, though it would have been no less tricky.

The Lotus and the Ferrari, each in its own way, are even more difficult. The Ferrari's bus-driver wheel position makes fast, accurate steering almost impossible. Our 308 suffered from a sticking throttle as well. At low speeds there is strong understeer, melting away to a nervous, on-edge feeling in turns above 100 mph, probably because of aerodynamic lifting of the nose.

The Lotus is a purveyor of odd and usually unpleasant messages to its driver. As cornering forces build, its steering gets light, even in the face of clear understeer. Turning into corners is nasty: after an initial steer in, you must quickly steer out to keep the tail from coming around. In the undulating, high-speed sweeper at Willow, the car was so darty that the driver would never again take it flat out after the first lap. Both the Lotus and the Ferrari are fast cars, but you have to grit your teeth to get the most out of them.

Surprisingly, the 911 was within a tenth of a second of the Ferrari, and it exhibited no trauma at all. It has fantastic steering: the forces build in direct proportion to cor-

nering force. You can feel what you're doing in this car: it sends clear messages to its driver. The big one is, "Don't get off the power in turns." If you pay attention, you can get around very quickly in this car and have a great time doing it.

---

Racetrack and cone courses are fine for satisfying curiosity and investigating handling off the beaten path, but what matters most is an automobile's real-life performance. What follows is as real as we can make it, the straight scoop from our Angeles Forest Highway logbooks.

### Lotus Esprit Turbo
### Gorgeous to Look at, but . . .

This car could be rechristened the Lotus Enigma. We all know what knee-high, midengined two-seaters with Lotus nameplates are supposed to be good for. Blurring the scenery, that's what. Making your sports-car-lovin' heart beat fast.

| | | roadholding, 300-ft-diameter skidpad, g | | maneuverability | | braking into a corner, mph | accelerating out of a corner, mph | SCCA Pro Solo gymkhana, mph | Willow Springs | | objective test summary* |
|---|---|---|---|---|---|---|---|---|---|---|---|
| | | smooth | bumpy | 900-ft slalom, mph | lane change, mph | | | | Turn Five speed, mph | lap average, mph | |
| | AUDI QUATTRO | 0.77 | 0.77 | 57.6 | 49.8 | 51.0 | 41.9 | 24.6 | 72.8 | 82.8 | 29 |
| | FERRARI QUATTROVALVOLE | 0.81 | 0.80 | 61.3 | 51.8 | 51.9 | 42.7 | 25.1 | 73.2 | 86.8 | 55 |
| | HONDA PRELUDE | 0.75 | 0.75 | 59.9 | 52.4 | 50.7 | 39.2 | 24.4 | 73.9 | 78.8 | 23 |
| | LOTUS ESPRIT TURBO | 0.81 | 0.82 | 59.0 | 58.1 | 53.0 | 40.0 | 25.2 | 72.4 | 87.1 | 55 |
| | PORSCHE 911 CARRERA | 0.84 | 0.82 | 59.8 | 53.7 | 51.7 | 42.1 | 25.6 | 74.2 | 86.7 | 61 |
| | PORSCHE 928S | 0.76 | 0.76 | 58.2 | 50.7 | 51.2 | 40.4 | 24.5 | 70.1 | 85.1 | 26 |
| | PORSCHE 944 | 0.82 | 0.81 | 61.4 | 53.3 | 51.9 | 41.6 | 25.1 | 72.4 | 84.4 | 52 |
| | TOYOTA CELICA SUPRA | 0.79 | 0.78 | 59.2 | 54.6 | 51.0 | 39.9 | 24.5 | 70.1 | 81.5 | 31 |

*Total points earned in nine track tests (8 = best, 1 = worst).      Maximum possible: 72

C/D Test Results

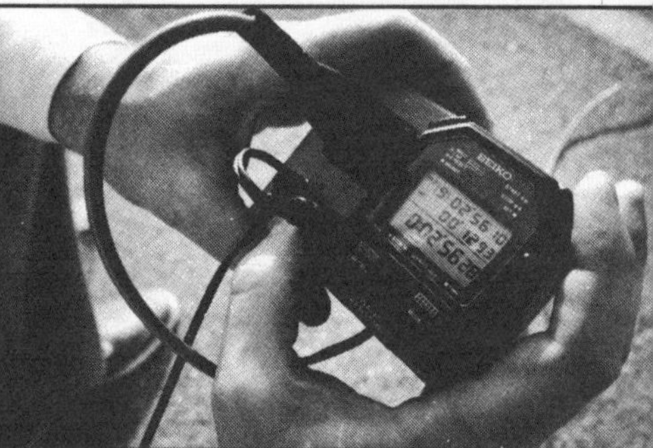

So it comes as a surprise that our testers ended up saying, "Great ride" (Sherman) and "Ride is wonderful on this choppy mountain road" (Csere). And lambasting its handling: "At seven- or eight-tenths, it comes unglued, unpredictable" (Sherman). "Start pushing it the way its looks and power say you ought·to, and the damn thing starts flopping around like a dying fish" (Lyons). "Twitchy and ornery" (Bedard). "A snap spin is probably in the cards for a beginner" (Ceppos).

Lotus spokesmen have told us of their goals for this car. They wanted a high-speed express, something in which a well-heeled European could jet across the Continent at triple-digit speeds in the company of one friend and minimal luggage. It had to be fast *and* comfortable, because a guy spending this much money doesn't want to be beaten up by his own machine.

The compromise for comfort is what produces the enigma, we think. A car so low and cramped inside is never going to be comfortable. "Shoes over size ten begin to get tangled in the pedals" (Bedard). "The shifter falls readily to elbow" (Ceppos). "Gives me a sit-in-a-hole feeling" (Sherman). Lotus did, however, soften the suspension: soft springs, soft shocks, soft bushings, particularly in the rear. The result is a wow-wee looker that rides well.

In routine freeway driving, or trolling on Sunset Boulevard, this works out fine (assuming you like the Esprit Turbo's appearance). The steering is light and sharp on center—"the classic manual-steering feel," according to one tester. "The gearbox is a delight around town," said another. "Engine is splendid once the boost comes up," said yet another.

Yet the Esprit always reminds one that it's not really constructed for comfort and convenience. The wide windshield pillars intrude on your vision, the wheel is so high you're forced into a praying-mantis position, and the seat shape keeps trying to submarine you under the belts. Of course, none of these intrusions are too surprising. In a Lotus, *everything* is subjugated to the one, true quality of a sports car, and that's handling.

Well, not this time. A Lotus engineer explained to us that ride is very important in Europe and the Esprit's ride has been substantially improved by the use of soft bushings at the forward ends of the rear-suspension trailing arms. He concedes that this does produce a measurable amount of deflection steer in back, but European drivers don't seem to mind.

We find ourselves minding terribly much. To a man, the testers disapproved of the Esprit's behavior when driven hard. And if you can't drive hard in a Lotus, what's it good for?

### Ferrari 308 Quattrovalvole
#### A Fine Mistress, a Poor Wife

Ferrari North America failed to deliver a test car representative of current production. Left to our own foraging, we turned up something at least as interesting: a low-mileage, gray-market example certified by an independent lab. It had Michelin TRX tires, as 308s had for years until they were replaced by Goodyear NCTs for 1984. Except for a few details such as a sticking throttle and a steering-wheel position higher and flatter than we recall in any other 308, it was a fine example of the breed. The engine was especially strong and zingy; the car felt light and frisky. In other words, a car like this could give Ferrari a good name in this country.

The testers' comments seemed unanimous. Ceppos could have been speaking for the entire corps when he noted, "Good brakes, intoxicating engine sound, stable low-speed understeer, awful driving position." The 308 Ferrari probably fits someone on this planet, but he doesn't work here. Sherman's knuckles rubbed the instrument panel at the top of the steering wheel. Bedard could barely reach that far when his seat was set at the proper distance from the pedals. All complained about the hot spot reflected into the windshield from the top of the instrument panel. Nobody cared much for the seat, finding it thin and hard and not particularly supportive of any part of the anatomy. Sherman observed that the instruments and the switches seemed to be casually sprinkled around the cockpit.

Typical of Ferrari practice is the friction in the steering. "The wheel can be turned and released, and it won't return to straight ahead, not even close," said Csere. "About half a turn from center the friction increases greatly; then toward three-quarters of a turn it drops way off to less than on-center," added Lyons. "Weird non-linearities as you turn away from center," confirmed Bedard. Sherman observed "occasional giant kickbacks in the wheel, particularly over low-speed bumps." Without the friction, the kicks would be worse.

The Quattrovalvole's suspension is highly damped, not moving much over the road yet reasonably soft in ride. Road adhesion is conspicuously good. In fast mountain driving, the Ferrari gets around with little slipping and sliding. "During throttle lift-off at my cornering speed, there is only a mild reduction of understeer—no serious threat of the tail stepping out," said Lyons. "Very predictable with lots of understeer, but it never really plows," said Csere.

The consensus is that the Ferrari is good fun when you feel like sporting around but a bit tight in the seams and pointy in the toes for everyday wear. And if you pay attention to business when the engine noise gets loud, probably the only trouble it will get you into will be with the cops.

## Porsche 911 Carrera
### Hanging In There

Strong feelings and Porsches go hand in hand. You won't find a more fiercely loyal owner body than Porschephiles, and our experience with the Carrera was likewise powerful. The consensus on the latest rendition of the ageless 911 is a mix of bubbling enthusiasm for its virtues, tempered with a healthy respect for its vices. "This is not a subtle car," said Csere.

The Carrera proved to be a triumph of painstaking development over antiquated design. The Son-of-Beetle rear-engine layout with its tail-heavy weight bias shouldn't work this well, but here is the Carrera, in its twentieth year, hanging in there with the best of them. On the test track our Carrera—outfitted with the limited-edition 930 Turbo bodywork, brakes, suspension, and wide wheels—blew the rest of the group away with the highest composite score. On the racecourse it was fast enough to turn in the third-fastest lap time and the quickest Turn Five speed.

In the acid test of street driving, the results weren't nearly as lopsided, but the reviews were never neutral. On the plus side, just about everyone agreed that the Carrera gives you a first-class office for your driving business. "Excellent seats, steering wheel, and ergonomics," commented Sherman. "The driving position is good," added Csere. About the only shortcoming is the near impossibility of heel-and-toeing, but since the Carrera's pedals are to some extent adjustable, it shouldn't be too hard to help this disability.

The Carrera's greatest gift, our merry band of testers agreed, is its forthrightness. In hard driving, "understeer and oversteer are honestly telegraphed," wrote Griffin. Bedard praised the steering: "No nervous corrections are required. Just wind it into the apex and then unwind." Griffin summed up the commendations: "I love it because it says, 'Here's the deal.' Sensational brakes, minimal dive, and tight steering are the saviors of this layout."

But there's another, darker side to the Carrera character. "It makes my palms sweat," wrote Ceppos. "All the multifarious messages," said Lyons, "add up to 'Don't tread on me.' "

What they're talking about is the 911 series' well-known disdain for lift-throttle cornering. Snap the throttle closed in the midst of an acrobatic maneuver, and the Carrera's tail will do a heart-fluttering step sideways. Happily, the Carrera is very stable when cornering hard on the throttle or on the brakes, but never getting caught out takes stern concentration indeed. "This car requires a deliberate driving style," noted Csere.

"It makes my palms sweat, too," wrote Bedard, but he wasn't specifically referring to the Carrera's handling. "How do they make a car with no weight up front steer so heavy?"

Ultimately, though, it was the level of mental, not physical, effort demanded by the Carrera that kept it from climbing higher in the pecking order. "This one will do for crazy fun," Sherman observed, "but you have to pay so much attention that it makes me wonder if it's really worth it."

## Porsche 928S
### Heavyweight Champion

The Porsche 928S was the widest, heaviest, and most powerful car in the group, and these characteristics created a sense of massiveness that dominated most of our impressions. Everyone noted its unswerving straight-line stability and self-centering steering feel. "Excellent tracking. Feels almost as if on rails," Csere observed. Ceppos found the 928 to have the "closest thing to manual steering I've ever felt in a power-steering car."

Its size also allowed plenty of room for comfortable and ergonomically correct interior accommodations. "The adjustable dash pod and steering wheel make it easy for me to get a perfect driving position," noted Griffin. Sherman fully agreed: "Perfect ergonomics, comfortable, hard to improve on. Excellent seat, visibility, transmission, and steering wheel."

But in the twisties, the 928's bulk did not serve it so well. Lyons called it "an overly pompous, lard-assed German—ponderous, clunky, clumsy, and overweight. This car is not happy doing this."

Griffin concurred: "The 928 never entirely loses its impression of bigness, of being a monster mutha on the roll—high, wide, and handsome." Bedard called it "big and cushy and isolated—very quick for its avoirdupois."

Much of this ponderous feeling came from the same high-effort, strongly damped steering that contributes so much to the 928's straight-line stability. Unfortunately, the heavy steering suggests a lack of agility. Several drivers felt that the high centering effort muddled the messages coming from the front tires. Lyons described it as "a swaddled feeling. Driving the 928 is like wielding a golf club or a tennis racket with oven mitts on."

However, the 928 can react very capably when given the proper control inputs. Ceppos said it provides "my idea of fabulous handling. This car's seat-of-the-pants feel is superb. The tires give you lots of warning, and you can work both ends of the car as you see fit. Fantastic."

Others, though, found a bit more oversteer than they liked. Griffin criticized its "tendency to hang its tail out like wash on the line, and with very little provocation." Bedard also found that it tended to "hang its tail out during transitions."

Yet everyone found an unbreakable, confidence-inspiring quality in the 928. Bumps, holes, and differing road surfaces never upset its ride or equilibrium. Lyons described it best: "Little complaints are totally overwhelmed by the tremendous sensation of competence this car exudes."

That's high praise for any car, but not necessarily the key to great handling. Great-handling cars should want to be driven hard and should urge their drivers to use their capabilities. The 928S has sufficient reserve within its 3440-pound soul to answer nearly any demand, but it prefers that you not ask too much too often.

## Toyota Celica Supra
### Japan Ascending

It's been four model years now that America's number-one importer has been claiming this car has "the right stuff." In most respects we agree. We generally like the Supra, and we were pleased with how well it held its own against the tough opposition it faced in the Angeles Forest.

The fine twin-cam 2.8-liter six remains a large factor in the Supra's basic appeal. In fact, as one tester noted, "the best part of the car's handling may be its *engine.* Always there, always willing, so torquey it almost never asks for a shift. Quiet and smooth, too." In Japanese fashion, the clutch and gearshift operations are so easy and pleasant they make the whole vehicle seem "user-friendly."

The famous Supra driver's seat struck most of us as excellent; the subjective ballot shows that we voted it the best in the test. Everyone liked the overall cockpit ergonomics as well, though some felt the steering wheel could be improved on (wanted: some place to put your thumbs), and one driver thought it high time Toyota rearranged the pedals for easier heel-and-toe operation.

Steering feel is not this car's finest point. Nearly everyone noted a numbness in the on-center position while cruising the freeways during the test. The front tires also had a mild tendency to follow longitudinal rain grooves. Off-center, the dead zone was replaced by a firmness that generated a more satisfying sensation; yet when cornering speeds were really cranked up, most wished for more feeling and feedback. "You steer purely by wheel angle," complained Bedard. "I found myself making unnecessary right-lefts in turns, just because I didn't have the right information from the steering."

On the lurch-provoking mountain roads, some testers would have appreciated a lower chassis stance and perhaps stiffer anti-sway bars. Not that the Supra behaves badly in these conditions—it's hard to find *any* conditions that make the Supra behave badly—but it does start feeling a little flustered and breathless earlier than a few of its rivals. *In extremis* handling can get a bit wobbly.

That's not because of any lack of basic adhesion. The Potenzas work quite well, and the cornering limit is signaled by a progressive, comparatively mild understeer. This is a benign quality that is not spoiled by the equally progressive, basically mild step-out of the tail when you lift your foot. Very much like the highly controllable 944, and somewhat like the more infamous 911, the Supra can be nicely balanced through a bend with delicate throttle movements. That's fun and satisfying. "Quite easy to drive like a demon," noted Sherman. Sharper bumps, though, do induce a fleeting instability while the rear axle takes a set.

The Supra is a jack-of-all-trades sort of car. It doesn't do any specific thing as well as some other cars, it lacks some final degree of the overall refinement offered by a few in our test, but it does everything across the automotive spectrum well enough to provide genuine driving satisfaction. The fact that the Supra costs less than half as much as most of the machinery in this test doesn't hurt its case, either.

## Audi Quattro Turbo
### All-Wheel Drive, No Jive

This four-wheeling supersedan—considerably improved for '84—was preceded by an imposing reputation. In many performance areas we were not disappointed.

Most of us, though, had trouble with the seat. Griffin compared the Audi's front bucket to an orange crate; Ceppos said the seatback seemed to be "ballooning"; others simply complained of inadequate lateral support. Sherman did note that the thigh support was good, and he called the steering wheel "ideal."

In terms of the steering itself, most testers declared this another car with an on-center dead zone and a lack of feel away from center. Sherman called it too slow. Bedard reported: "There is no effort build-up to coordinate with cornering force. You end up steering by wheel angle only."

The unhappy steering contributed to the marks the Audi lost on the freeway segments of our test because of the uneasy feeling it generated while changing lanes. Minor steering inputs seemed to cause an excessive amount of body roll, and the car required constant attention to hold a line. In view of the car's tendency to pitch and wallow, the ride quality struck some as inordinately harsh.

The Quattro's good side came out on the mountain roads. "Hard to find serious fault," Csere noted. "A lot of adhesion, great stability, good brakes, excellent turn-in capability, and very good manners." Ceppos said it didn't have the cool, calm confidence of, say, the 928, "but you can boogie with it." Lyons likened the feel to that of the Prelude, where he felt at home and able to go fast immediately, but with acceleration and adhesion added. "More car here than I really need or can use," he went on. "But, unlike some of those performance cars that give me the same impression, the Quattro Turbo contrives to do it in a friendly way. It doesn't intimidate. It gives me messages that keep assuring me, 'Hey, sure, stay with it, buddy. You can handle it.'"

The road surfaces we encountered did not permit us to test all the potential advantages of four-wheel drive, but it was interesting to play with the three differential-locking options on the twistiest mountain sections. The differences were subtle: with the center and rear diffs locked, there was a comfortable trace of understeer near the limit; with both lockers free (the normal dry-pavement mode), the car seemed comparatively neutral, and there was also a trifle more tail-out available by lifting off the gas pedal.

The turbocharged five-cylinder engine offered plenty of power, even at mile-high altitude, and also contributed a strong-sounding, almost strumming note that kept us impressed with how much machinery was in our hands.

An interesting car, impressive and definitely worth more exploration. The testers agreed that there is a lot of potential left in the design.

## Honda Prelude
### Front-Drive Fans, Unite!

We're flying at very high altitude here. The Honda Prelude is the only car in this test with front-wheel drive, it's by far the least expensive, and it has proved more than a match for every car but one.

The Prelude didn't win outright, but it is so good that it must be considered a true ace. The piercing of California's mountains by this feisty little rice rocket quickly escalated our studied commentary into a crescendo of glad tidings.

"The variable-assist power steering is very light at low speeds and then progresses to fine firmness at high speeds," said Griffin. "Tracks better at 80 than at 60." Csere found the steering "good, tight, with no lost motion. But effort is low. Wheel doesn't tend to return to center."

Sherman had high praise for the Prelude: "Great car. Excellent ergonomics. Easy to drive. Good directional stability, but a touch of wander. Quick and responsive off-center. Good linearity."

Bedard liked "this low-effort steering. Sharp enough that you don't have to worry it—just go on instinct. With the Porsches, you are always conscious of steering." Ceppos, however, noted that, although the Prelude "doesn't steer like the cheapest car in the group, it doesn't quite have the 944's laser-accurate tracking, either."

Lyons took a broad view: "No wonder these are so popular. It's as pretty in its functioning as it is to look at. If you don't ask too much, the Prelude is steady, comfortable, willing."

Then we latched onto the mountains.

"Absolutely no treachery," said Bedard. "It telegraphs every move. A novice could be warned back from disaster." Ceppos termed the Prelude "one of the nicest-driving all-around cars ever. Nothing tricky about going quickly, but an expert can play the controls to make it do just about anything." "Dead pedal okay," said Sherman, but added, "Wheel needs thumb spokes."

"Good grip," wrote Csere, "but, more important, the Prelude's extremely secure. Understeer can be killed instantly by just lifting. Brakes strong, progressive. Can be braked and turned at the same time. Generally extremely forgiving—and fast!!!"

"I love it," said Griffin. "A stupendous value. Sails through this as if there were no obstacles. Far and away the most stable. I'd

| Subjective Scores* | | ergonomics |
|---|---|---|
| | AUDI QUATTRO | 34 |
| | FERRARI QUATTROVALVOLE | 8 |
| | HONDA PRELUDE | 36 |
| | LOTUS ESPRIT TURBO | 6 |
| | PORSCHE 911 CARRERA | 24 |
| | PORSCHE 928S | 45 |
| | PORSCHE 944 | 39 |
| | TOYOTA CELICA SUPRA | 37 |

buy one like a shot and wouldn't change a thing—well, another 30 hp under the hood would be grins."

Lyons: "You follow the Lotus in this car, see it wobble, and wonder why. You can corner with the 928 and the 911. Power lets them pull ahead, but not by much."

In our book, the superb Honda Prelude is the second-best-handling foreign car any amount of money can buy.

### Porsche 944
### All Hail the Proud Victor

The Porsche 944 is the best-handling imported car. It will easily zig past the near great and easily zag with the truly great in their brightest moments of glory.

Wild-eyed Porschephiles will tell you a win by one of their own was predictable. Haters of the marque will say it's bogus because *C/D* inevitably pays homage to all Teutonic machines. But the 944 won overall simply because, in its remarkably complex but also wonderfully integrated way, it has no all-around equal.

"Power steering light but accurate," said Griffin. "Much road noise—how uncivilized—but steering and chassis ignore pavement trivia."

Csere: "Fine steering, no lost motion."

Jazzman Ceppos: "Very refined except for Gene Krupa snare drums over every tar strip and Botts dot. But tracks like a bullet."

"Force buildup just off center is much greater than in manual-steering 911," said Bedard. "Prefer the 944. Great steering."

"No steering kick," said Sherman. "Ergonomics great, though wheel is low. All systems go."

Lyons found it "hard to believe the 944 is built by the same company that makes the 911—feels modern but inexpensive. Leans too much."

Sherman: "Confidence-inspiring. Easy to learn. Terrific turn-in, linearity, brakes, engine, dead pedal, ride and handling over bumps. Needs the better thigh retention of optional sport seat. Carves up corners the way you want. Great piece!"

"Miss grip of P7s with P6s," said Lyons, "and also iron-fist mechanical feel of 911. A *boulevardier* by comparison, but can play grip at front and back to heart's content. A benign chassis."

"Unlike our long-term 944 in Michigan, this one understeers too much," said Griffin. "As a whole, too civilized after 911, but lovely for full-time transport."

"Lots of grip and balance. Brakes soft in initial travel but quite good," said Csere.

"Well-rounded athlete," said Ceppos. "Not outstanding in any category, but fun. I could go for more stability in tail."

"Pure magic," said Bedard. "Tells you everything. Neither end lets go suddenly. Very important: excellent coordination between buildup of cornering force and steering effort and angle. A thoroughbred."

The 944 offers drama to go, but the blip of the driver's pulse rises only from the rush of emotional reassurance that comes with every new discovery of its sporting behavior.

A driver need not be a speed master to drown happily in the liquid responses of this car. This is a Porsche, to be loved for what it is as well as for what it can do. We have no doubts at all that the 944 is the best imported handler to be found in America.

---

Are you surprised by our findings? We certainly were. They make perfect sense when you stop to realize that automobile handling has indeed advanced by leaps and bounds during the past couple of years. The manufacturers inclined to rock back on their laurels and watch the world go by are bound to suffer, while the up-and-comers who eagerly exploit the latest tire and suspension technologies advance their causes tremendously. Porsche is in the thick of things with the best-handling import in all the land, and a couple of mid-pack finishers to boot. The Japanese have managed to bracket the prestigious Audi Quattro with the Honda Prelude's fantastic second-overall finish and an equally remarkable fourth overall by the Toyota Supra. You'll notice a rather sharp drop in the subjective scores on the way down to the Formula 1 crowd—Ferrari and Lotus. The 308 Quattrovalvole and the Esprit Turbo are elderly designs, mid-engined or no, and it's clear that both their interior accommodations and their steering and suspension components are in need of serious attention.

Now that we've determined the best-handling domestic-made car and the best-handling import, there's only one thing left to do. You guessed it: Chevrolet Camaro Z28 versus Porsche 944 at forty paces. As soon as our g-suits are back from the dry cleaner's, we'll be back on the road for the grand finale.

*—Patrick Bedard, Rich Ceppos, Csaba Csere, Larry Griffin, Pete Lyons, and Don Sherman*

| driver's seat | comfort | steering sensitivity | | steering response | directional stability | adhesion | | ride | cornering while | | THE ANSWER** |
| | | on-center | off-center | | | smooth | bumpy | | braking | accelerating | |
|---|---|---|---|---|---|---|---|---|---|---|---|
| 22 | 30 | 33 | 28 | 29 | 33 | 43 | 41 | 36 | 44 | 40 | 37 |
| 17 | 17 | 14 | 24 | 25 | 34 | 36 | 30 | 26 | 30 | 36 | 22 |
| 34 | 43 | 40 | 40 | 42 | 36 | 33 | 36 | 38 | 44 | 41 | 41 |
| 8 | 9 | 26 | 21 | 19 | 19 | 33 | 29 | 35 | 19 | 26 | 9 |
| 45 | 32 | 36 | 40 | 39 | 30 | 39 | 36 | 24 | 22 | 39 | 29 |
| 42 | 44 | 35 | 32 | 38 | 33 | 36 | 34 | 38 | 26 | 38 | 29 |
| 38 | 35 | 42 | 46 | 45 | 39 | 41 | 40 | 34 | 40 | 40 | 42 |
| 46 | 41 | 34 | 31 | 31 | 34 | 34 | 32 | 38 | 34 | 36 | 30 |

*Total scores awarded by six judges (8 = best, 1 = worst). Maximum possible: 48    **Overall rating

# Best-Handling Bottom Line

*Two Z28s, one Porsche 944, seven judges, many roads, but only one big winner.*

• In the final analysis, automobile handling is a very personal matter. At the test track, it's possible to take the nut behind the wheel out of the system. But in the real world, only a human being can separate good traits from bad ones.

So here we are, back on the road again, with real humans, to determine the best-handling car. Our stopwatches, accelerometers, and cone courses have done their duty; now it's up to Rich Ceppos, Csaba Csere, David E. Davis, Jr., Larry Griffin, Jean Lindamood, Don Sherman, and Brock Yates to sit in judgment. The significant seven have weighed the test-track and racetrack results against their real-world experiences to arrive at seven personal bottom lines. Simple addition of the seven votes will reveal our winner.

This is the title bout, the main event in our three-round search for handling excellence. A Camaro Z28 vanquished all domestic comers last May, and a Porsche 944 humbled all other import contenders in July. The two victors meet on these pages in a final confrontation. One will go forth our proven champion; the other, we hope, will benefit from the experience.

---

Although California's Angeles Crest Highway served us well in the two previous tests, we selected some of our favorite roads in Ohio for this third round. By now, the inhabitants and law-enforcement agents in and around New Guilford, Ohio, know us well: their State Highway 541 has measured the mettle of many of *C/D*'s road warriors. Our carefully selected stretch of asphalt slices through a whole county of Mother Nature at her feistiest. The beauty of the route is that civilization has barely scratched the surface here. The ride offers a full dose of hills and hollows, twists and turns, bumps and bends; the graders were apparently so busy clearing a path through the forest that they never quite got around to planing the earth flat, straight, and perfect. The locals seem to keep their cats and dogs collected and their various autos and implements safe at home whenever we come to play, so 541 is hard to beat as a real-world test site.

This month's best-handling puzzle consists of four irregular and odd-sized pieces. Chunk number one is the biggest and most important component: what we learned on twisty, turny Route 541. Chunk two is a good deal smaller and easier to grasp: straight-road handling, or what we experienced in transit to and from New Guilford on Michigan and Ohio multilane highways. Chunk number three comes from the test track: the results of various carefully controlled roadholding and maneuverability experiments (see "*C/D* Test Results"). The final piece, a small one, consists of lap and segment times from a racetrack. We feel the closed-course data are useful cocktail-hour ammunition but hard to translate into meaningful terms when the belts are cinched and the engine is keen to run you down a real-life road. For this reason; the racetrack results received low emphasis.

Once all the data were in, our human computers found a comfortable spot under the New Guilford Church of Christ's best shade tree, then sat down to fume and fidget for a while. Some of the judges made



PHOTOGRAPHY BY AARON KILEY

up their minds immediately; others went to work on elaborate charts and formal briefs justifying their decisions. A few last-minute test runs were followed by more deep deliberation.

Eventually, we called a halt to the ruminations and ordered pens to paper. When the verdict was in, the Porsche 944 beat the Camaro Z28 five votes to two.

This clear preference for the 944 is misleading in one sense: the net difference in fun-to-drive and overall handling competence between the two cars is far smaller than the five-to-two ratio suggests. In other words, this opus is far from over. We've revealed The Answer early to spare you the suspenseful buildup, but the story behind the Porsche 944's win can come only from the testers' logbooks.

First, a couple of vital statistics: At 3380 pounds, the Camaro Z28s under examination (we brought two on the trip, each having a high-output engine and a five-speed transmission) outweighed the Porsche 944 by a significant 560 pounds. In addition, the Z28 rides on a wheelbase that's longer by 6.5 inches, and it hauls around 5.8 inches more width and 17.8 inches more length than the Porsche. The extra bulk is partially offset by higher-aspect-ratio tires (with the same section width and rim diameter as the Porsche's tires) and a good deal more power under the hood. Nevertheless, the Camaro is always hauling around an inertia handicap, and this can only hurt its handling prowess. David E. Davis, Jr., summed up the relative heft of the two cars with the following insight: "I don't seem able to toss the Camaro around as lightheartedly as the Porsche. All the Z28's controls are accurate and positive, but each one transmits that feeling of heavy effort overcoming great weight. It's all push and pull."

The first leg of the journey included a bad section of four-lane highway that offered an excellent test of ride comfort. You might guess that the Camaro's extra mass would be helpful for hammering the expansion joints back into the pavement, but this really wasn't the case. Sherman report-ed lots of "boom-boom" noise in the Camaro over the rotten pavement and the nastier expansion strips but added that the solid-feeling structure did at least do a decent job of attenuating the punishment from the road. The Porsche came to this test in base-suspension trim (extra-cost anti-sway bars, heavy-duty shock absorbers, and lower-profile tires are available from the factory), and this may well have been an advantage on the hard-knock highway. Ceppos, Griffin, and Davis all described the 944 as "supple" over the rough pavement, our fearless leader adding that the Camaro's deep, plunging vertical motions were not present in the 944.

Although the Porsche's combination of mass, spring rate, and bump-stop stiffness seemed to work better over the battered and broken stretches of U.S. 23 in Michigan, the Z28 came into its own once we encountered smoother roads in Ohio. While the Camaros were cruising quietly and contentedly, the 944 was generating mild complaints. According to Yates, "The 944's tires and suspension make some unseemly rumbles at 70 to 80 mph."

On-center stability was excellent in both brands, though the Z28 seemed to show a slight advantage in tracking. It demonstrated no tendency to nibble out of line over changes in road camber or when encountering longitudinal ridges in the pavement. Csere was happy with its on-center steering feel, reporting "no lost motion as the wheel passes through straight-ahead." Griffin

and others noticed that the Porsche's steering was "active" over minor road imperfections, though this was considered more of an irritant than a major concern. Sherman and Lindamood concurred that the Porsche required more course corrections on the highway, but they felt that this handicap was offset by on-center steering precision that was superior to the Z28's.

The distinctions between the Porsche and the Camaros came into clearer focus once we left the four-lane and set off at a good clip down the Ohio byways. Davis recorded heartfelt praise for the Porsche: "Truly quick and agile on this stretch. Very nice on chatter bumps. Always stays pointed. Steering is wonderfully light on transitions. The car is quite stable under heavy braking in the corners and on rough surfaces."

Lindamood also waxed ecstatic over the 944: "Feels so fine in turns. Steering feels quite secure. Rear end slides out slowly and is easily reeled in. The chassis is not disturbed by rough road."

Rich Ceppos, however, discovered a few flies in the ointment: "The 944 feels soft and just a little uncoordinated under pressure. The bobbing tail, twitching wheel, and so-so directional control on these rough-edged roads mean you have to mind it more. I think it is as fast as the Camaro, just not as stable."

Sherman chimed in with a mixed review for the 944 Porsche: "Good steering precision, use-anytime brakes, super engine and gearing. The bad news is that the back end goes away at times, particularly over sharp crests. One must also learn not to pitch the car in toward the apex too aggressively, or rear adhesion can be lost."

Yates was nearly all smiles for the 944: "Yes, Bucky, there is a substitute for cubic inches. Great ergonomics, great balance among braking, suspension, and available power. Very controllable under all reasonable conditions but surprisingly vulnerable over bumps. The steering needs too many little corrections on the straights."

Observed Griffin: "The 944 is not viceless, but its pluses overwhelm its minuses. I'm nuts about the steering effort, the sureness of the brakes, and the precise tidiness of the chassis. Yes, it moves right and left in a straight line from time to time, but 98 percent of the time it's telling me what I want to know and reminding me why I want to know it."

Csere was also enthusiastic about the Porsche's road manners: "It's tied to the road better than the Camaro and is thrown less by the bumps. Not too tail-happy—and when the tail does come out, it isn't a problem. The brakes are far better than the Camaro's and much more progressive. They help make the car more forgiving." Csere did note some negatives: "The Porsche simply doesn't turn in as crisply as the Camaro. And there are stability problems. The tires follow minor road imperfections. There is steering-wheel kick, both on the straightaways and in the turns."

The Z28 logbooks were just as full of praise and protest. Nearly everyone voiced a negative reaction to the brakes. Csere called the system "nonlinear. There's a high initial resistance to pedal pressure that delivers no stopping power whatsoever." Ceppos termed the brake feel "mashed potatoes." Sherman said, "The brakes are slow to respond. You must get into them early and then not use them too hard to avoid the onset of fade."

On the other hand, nearly everyone praised the Z28's tight grip on the road. Sherman exclaimed: "You can't possibly use all the traction. This car is permanently stuck. Even if you're foolish enough to slip into the marbles, the tail doesn't drift out of line." Csere added: "In the Z28, I barely slid a tire anywhere. The steering is excellent—direct, linear, confidence-in-

| Vital Statistics | | price, base/as tested | powertrain | engine/ transmission | suspension | | curb weight, lbs |
|---|---|---|---|---|---|---|---|
| | | | | | front | rear | |
| | CHEVROLET CAMARO Z28 | $11,022/$14,800 | front-engine, rear-drive | 5.0-liter V-8/ 5-sp manual | ind, MacPherson strut, coil springs, anti-sway bar | rigid axle, 2 trailing links, torque arm, Panhard rod, coil springs, anti-sway bar | 3380 |
| | PORSCHE 944 | $21,440/$23,325 | front-engine, rear-drive | 2.5-liter 4-in-line/ 5-sp manual | ind, MacPherson strut, coil springs, anti-sway bar | ind, semi-trailing arm, coil springs | 2820 |

spiring. This car is definitely easier than the Porsche to drive fast. Tail-out worries don't exist." Ceppos agreed: "This car has so much grip that I can't probe the limits on this road. It really sucks up the bumps and points and steers like a champ."

Several judges were less enthusiastic. Lindamood felt that the limits were higher and more secure feeling in the 944. Yates felt that the Z28 would be more at home on the fast, flat, open bends of west Texas or New Mexico. He called the Camaro "a very sophisticated American car but not much else. It might be very quick on a racetrack, but it's a bit of a chore in the real world." Griffin opined that "the Z28 isn't in the same league as the 944. The bumps, dips, swoops, and camber changes leave the Z28 floating where the 944 stays stuck."

Davis felt that "driving the Camaro 70 to 75 on these roads is like driving the best sedan you ever drove. It is wonderful, but it is big! It would be much easier to hurry the Porsche through here with one's wife riding shotgun. The lady would quickly object to the beating she was taking in the Camaro." Yates agreed that "the Z28 does nothing overtly offensive and it is a pleasant car, but these humpy, bumpy conditions suit the 944 better."

The Z28 was a model of stability. On the

| weight distribution, % F/R | wheel-base, in | tires |
| --- | --- | --- |
| 56.8/43.2 | 101.0 | Goodyear Eagle GT, P215/65R-15 |
| 48.9/51.1 | 94.5 | Dunlop SP Sport Super D4, 215/60VR-15 |

# A Feel for Handling

*Handling and roadholding are two different things, and the numbers tell only part of the story.*

• This is the third and final story we've done, as, by a process of elimination, we've zeroed in on the best-handling car available to American drivers. As the series has developed, it has become increasingly clear that a large and vocal minority of our readers is confused about the difference between handling and roadholding, and suspicious of our methods for evaluating one as opposed to the other.

Some of our correspondents have accused us of being—horror of horrors—*subjective*. Well of course we're subjective. Everything about automobile handling is subjective. Roadholding can be judged through precise measurements of the car's dynamic performance against the road surface, and is thus objective. "Roadholding" is an expression that describes exactly what a car does, on the skidpad, or going through a given corner on a certain circuit. "Handling" primarily describes how the car *feels* when it's doing those things, and, secondarily, when it's doing everything else it does in its workaday life. Roadholding is what it does. Handling is how it feels. Roadholding is one component of handling. Handling is more—handling describes our entire physical relationship with a vehicle in motion.

Our definition of good handling is not, as some readers have suggested, mere "ease of driving," nor do we regard handling as good when a car exhibits some nasty dynamic quirk that makes it exciting, like a sudden transition from understeer to oversteer. Quirky, exciting cars may turn out to be great production racers, or Solo II contenders, but they are not necessarily what one wants for a fast ride to grandmother's house on wet and winding roads. Race drivers are paid to tame cars that are savage beasts. The owner of a production car expects the engineers to have tamed it before he makes his purchase. Larry Griffin once asked Danny Ongais—maybe the bravest race driver ever to draw breath—to describe his technique for fast cornering in his personal Porsche 930 Turbo, and Ongais answered, "I don't go around corners fast in my Turbo."

Years ago, I was part of a group evaluating some cars at the old Packard proving grounds in Romeo, Michigan. Another member of that group was John Fitch, former member of the Mercedes-Benz factory team and gentleman inventor, who told me that one of the things the great engineer Rudolf Uhlenhaut used to stress to the Mercedes team drivers was the fact that the fastest car was not always the most pleasant to drive. His point was that racing drivers should think twice about insisting upon setups that "felt" right; that what really counted was measurable performance improvements.

This is almost diametrically opposed to the way production cars are set up by their engineers. First and foremost, the production car must feel good to its driver. It must "handle" well. Ultimate roadholding—even in a very high-performance sporting machine—isn't going to be all that important to the average owner, because he really won't be operating the car at its limit of adhesion on the public roads. Far more important is the need for a car that *feels* right, that sends the right signals, that seems to want to go where it's pointed—in short, a car that handles well.

Generally speaking, *Car and Driver* is less interested in the pure numbers—in this case, the numbers pertaining to roadholding—than in overall performance, expressed in more human terms. The 1984 Corvette provides an excellent example: it is fast and it generates terrific numbers on the skidpad, or anyplace else for that matter, but it really doesn't handle all that well. It feels like a race car. Insiders at General Motors refer to it, fondly, as a brute, and that's what it is. It is undeniably true that one can drive wood screws in with a sledgehammer, but an electric screwdriver does the job ever so much more elegantly.

Now, in the case at hand, that of the Chevrolet Camaro versus the Porsche 944, we saw the Chevrolet product generate remarkable numbers, yet we picked the Porsche as the best-handling car. The Camaro was, and is, the best-handling American car in our experience. Its roadholding, among American cars, is second only to that of the Corvette. But a majority of the test drivers opted for the Porsche in the broader regions of handling. We drove both contenders over a variety of roads, in every mode from Little Old Lady to flat out, and it was our reasoned judgment that the 944 was the better-handling car. That's how it felt to us.

*—David E. Davis, Jr.*

very manageable runner that responds kindly to just about any driving style. If anything, it is more telegraphic on the track than on the street. The brakes work surprisingly well and are quite easy to modulate. The Porsche demands more care in the corners, because of its tendency to lift-throttle oversteer. Whereas the 944 pitches and twitches over bumps, the Camaro is always steady-as-she-goes—much less nervous." The final click of the stopwatches told the rest of the story: the Z28 was 1.5 mph quicker in lap average and 0.8 mph quicker through the tight left-right combination leading onto the start/finish straightaway. The Z28 eats sharp transitions of this type for breakfast, while the 944 asks for a more deliberate touch at the steering wheel to keep all four tires within their traction limits.

Chevrolet earned one more Brownie point during this best-handling exercise by outfitting one of its test cars with a new, improved version of the Conteur seat. Strictly speaking, this was a 1985-model prototype part, not really pertinent to our production-car sample, but the logbooks bubbled with enthusiasm for the optional seat anyway. Alterations from the existing design include a deeper butt pocket, better padding over the side-bolster structure, and a new thigh-support section that adjusts fore and aft instead of up and down. It's a dramatic turnabout for the very seat that used to generate our most pointed criticism.

---

In the final tally, it's impossible to ignore the fact that two very different automobiles have climbed the mount for this final face-off. In Brock Yates's words, "It was the battle-ax versus the rapier." The Camaro Z28 is the fat-tired, torque-laden, modern-idiom muscle car. It's a wonderful improvement over any past Z28, and still the best there is in made-in-America machinery, but it's packing plenty of hardships in its portfolio: a quarter-ton too much weight, a floorpan that's ten-percent oversize, and brakes that won't operate in sync with the rest of the car.

The Porsche is by no means perfect, but the two most important handling ingredients are baked deep within its soul: a fine balance of all the components that make the car go, turn, and stop, and an unshakable dynamic poise. May the 944 have a long and fruitful reign as the best-handling automobile in all the land.

*—The Significant Seven*

rare occasion that its limit could be reached on the open road, the breakaway was inevitably a safe, slow understeer that was easily corrected. Several judges used this trait to good advantage in pushing the sporty Chevy harder and faster, while others interpreted the Z28's steady nature as a sure sign of sloth and sullenness.

Everyone agreed that the Porsche felt light on its feet, ever eager to dance left or right. Although it lacked the Z28's permanently planted feel, many judges felt this to be a positive characteristic, indicating superior agility. The Porsche was certainly a master at telegraphing its moves to the driver. Frequent slips and slides from the back of the car served as clear signals that some limit was fast approaching, and these clues were typically offered early enough that minor corrections were effective in rectifying the situation. The Porsche also did a superior job of speaking to its driver through the steering wheel. Steering effort increased in direct proportion to the cornering force, a trait sadly lacking in the Z28. A couple of the judges also liked the kick fed up through the Porsche's power rack-and-pinion steering gear, feeling that it was an essential ingredient for "athletic endeavors."

---

The Z28 may have failed to sway the majority when it came down to the final vote, but it did score several significant points. Track tests between the $23,000 Porsche and the $15,000 Chevrolet resulted in a dead heat. The Porsche bettered the Chevy in four of six cornering tests, but the Z28 trounced the 944 on the racetrack.

We ran the two contenders head to head at Nelson Ledges only a few days before the annual Longest Day event; the results are especially pertinent since Chevrolet and Porsche are mortal enemies on this track. What's more, the two have collectively won four of the five 24-hour races at Nelson Ledges. (Porsche prevailed with a 924 in 1981, with a 944 in 1982, and with a 944 Turbo this year, while Chevrolet won with a Z28 in 1983.)

Rich Ceppos, who handled all the racetrack driving, reported that "the Z28 is a

| | | roadholding, 300-ft-diameter skidpad, g | | maneuverability | | braking into a corner, mph | accelerating out of a corner, mph | Nelson Ledges | | objective test summary* |
|---|---|---|---|---|---|---|---|---|---|---|
| | | smooth | bumpy | 900-ft slalom, mph | lane change, mph | | | Turns Twelve and Thirteen, mph | lap average, mph | |
| | **CHEVROLET CAMARO Z28** | 0.81 | 0.77 | 60.9 | 55.1 | 52.0 | 41.3 | 44.1 | 84.3 | 12 |
| | **PORSCHE 944** | 0.82 | 0.81 | 61.4 | 53.3 | 51.9 | 41.6 | 43.3 | 82.8 | 12 |

C/D Test Results

*Total points earned in eight track tests (2 = better, 1 = poorer). Maximum possible: 16

# DON SHERMAN

*"You don't have to be bad to get better."*

---

• A handful of automotive journalists were recently summoned to lunch in Beverly Hills, California, to witness the detonation of a bomb of no mean proportions: Porsche was revamping its entire distribution system in the United States. Peter Schutz, the 53-year-old German-American who is president and chief executive officer of Dr. Ing. h.c. F. Porsche AG and also president of a new American holding company called Porsche Enterprises, punched the big red button while the waiters blithely served filet mignon.

Schutz had been rattling swords for months in Germany, and the news had already broken, two weeks before the press luncheon, that Porsche AG would sever its relationship with Volkswagen and Audi in the U.S. In spite of the warnings, surprise was the main course served by our hosts that day. It's hard to prepare yourself mentally for thermonuclear housecleaning.

Since the blast's epicenter was well removed from Beverly Hills, no fallout soiled the table linens. However, that might not have been the case had the journalists been invited to lunch with Herr Schutz the previous day when he broke the news to the Porsche+Audi dealer body. According to trade-paper reports published shortly after that fete in Reno, Nevada, the cries of anguish and the screams of financial pain were deafening. The stampede to the telephones—dealers rushing to energize their attorneys—looked like a reenactment of the evacuation of Poland. Even Schutz later admitted that the nearly universal refrain from his disenfranchised retail agents was, "You've just put us out of business."

Before you pack a charity box of sandwiches and secondhand clothing for your local Porsche+Audi dealer, it might be wise to ponder his plight. As of September 1, 1984, he will no longer be able to purchase Porsche automobiles from Volkswagen of America for retail distribution. He will, if he chooses to sign up for the new franchise offered by Porsche Cars North America, have the opportunity to carry on business pretty much as usual as a Porsche retailer, so it's not too likely he'll end up in the bread lines. The rub is that many of the Porsche dealers love the little gold mines they're now operating and want nothing to disturb their profit picture.

Peter Schutz's view of the balance sheet is understandably a bit different, and the announcement of his plan offers a rare look inside the Porsche factory's psyche. The move makes sense in context, because Schutz has systematically reversed many of the postures he found in place when he arrived in Stuttgart to take the helm late in 1980. The master plan at that time said the Porsche 911 would cease and desist after the 1982 model year. Schutz instead revitalized the rear-engined throwback and gave it a future. The marketing department, circa 1980, felt that the world would soon be hostile toward high-tech, high-priced sports cars because of energy and environmental concerns and suggested a minimal-investment strategy. Schutz disagreed, committed all the resources he could muster to a bigger and better Porsche, and is now reaping strong sales and unprecedented demand for his trouble. Likewise, the Porsche racing team was racked with paranoia, worried that the press would eventually show revulsion to the Stuttgart steamroller's success in motorsports. Schutz reacted to those notions by dusting off a 936 museum piece and approving an all-out preparation program for the 1981 24 Hours of Le Mans. Jacky Ickx and Derek Bell won going away, and the press kissed Porsche's feet, as usual. In 1982, the 924 was a dead player against the Japanese and American competition, and many dealers whined for a price rollback. Schutz chose the opposite tack: he legitimized the 924 by making it into the 944 and *raised* prices fifteen percent. Two years later, dealers still charge premiums for spots on their waiting lists.

Schutz has become one of the great decision makers in the car world, but it's clear that the fundamentals he learned as a mechanical engineer still guide his actions. The man began his career in the diesel-engine development section of the Caterpillar Tractor Company, and he has been a devout practitioner of the test-and-evaluation method ever since. The 911 Cabriolet, the Porsche 944, and the 1981 Le Mans project were all experiments of one sort or another, but they pale in comparison with pushing the big reorganization button. This is the acid test of them all: will the dealer body support Schutz's "you don't have to be bad to get better" theory?

In a word, no. The dealers as a group don't feel they need to get that much better. Sales are the best they've ever been, profits are rosy, and the customers are reasonably well satisfied.

What's more, automobile dealers in America are a powerful group. They may not look so united when they're bidding against one another with rebates and free lube jobs, but under fire, they close ranks to become a well-oiled war machine. It took the National Automobile Dealers Association, the American International Automobile Dealers Association, World-Wide Volkswagen, and a Porsche dealers' action committee about three weeks to file lawsuits against Schutz's plan, seeking *billions* of dollars in damages. The bomb was effectively back in Schutz's lap, and ticking.

His plan would have established 40 new Porsche Centers that would have been factory-owned stores. The centers would sell new Porsches at sticker price, refurbish used ones with factory-grade repair and repaint capabilities, offer 24-hour emergency service, administer a new lease program, and provide an electronic link to the Zuffenhausen factory for special orders. The dealers would have been downgraded to Porsche "agencies" and paid a greatly reduced fee to drum up sales. The dealers' lives would be simplified because they would have no inventory to finance (they would sell from a national pool of automobiles), but the ones we've spoken to felt their businesses would be destroyed by such a plan. Competition with the factory stores for sales and service would slice them off at the knees. Forty dealers were offered a chance to run the Porsche Centers, but only two signed on the dotted line.

Schutz had little choice but to slip the pin back into his grenade. In a letter to Porsche+Audi dealers, John A. Cook, the president of the new Porsche Cars North America, announced that "the Porsche Centers will *not* be established as originally planned, and that a retail distribution network, utilizing independent dealers, will be implemented." This is a new twist on the old give-and-take: the dealers gave nothing, and Peter Schutz took back his plan for the future.

Perhaps a new motto would be appropriate. Something like: "You don't have to be bad to get better, but if you're going to play God with the franchise system, don't expect Adam and Eve to sell your automobiles."

●

# Life with 944

*We'd gladly go another 30,000.*

● If ever a car was an immediate sales success, the Porsche 944 was. Overnight, the transformation of the 924 into the 944 changed the sales picture of the entry-level Porsche from slow-moving to hot. One reason was the replacement of the 924's rough and anemic 2.0-liter engine with a smooth and powerful 2.5-liter four derived from the 928 motor. Another was the addition of muscular fender flares housing vastly larger and stickier tires. Furthermore, the 944 was fully equipped with electric windows, air conditioning, aluminum wheels, and four-wheel disc brakes—yet at $18,900 it cost barely more than its predecessor. It also promised good fuel economy and had longer service intervals than the 924.

We fell for the new Porsche hook, line, and sinker and added the good *C/D* name to the long waiting list. Unfortunately, we were slower than a lot of you to jump on the bandwagon, so we didn't get our long-term car until March 1983, nearly a year after the introduction. By then the base price had climbed to $19,485.

In addition to the standard features, our car was equipped with power steering, a Blaupunkt Monterey AM/FM-stereo radio/cassette, the four-spoked, leather-covered steering wheel, a limited-slip differential, sport seats, "sapphire metallic" paint, the anti-sway-bar package (a rear bar and a larger front bar), and the classic five-spoked, forged-aluminum wheels (seven inches wide in front and eight inches in the rear, instead of sevens all around). The total came to $23,155—a lot of money, but not unreasonable for a loaded Porsche in this day and age.

At least the price bought us a great-running 944. Our long-term test vehicle was

actually a bit stronger than our original road-test car (*C/D*, May 1982), sprinting from 0 to 60 mph in just 7.4 seconds and climbing to a peak speed of 129 mph. The wider rear wheels and the anti-sway-bar package boosted the skidpad adhesion from 0.81 to 0.83 g.

Best of all, we could readily use this performance on the road. The 944's engine pulls well at all rpm, and its smoothness encourages frequent runs to the redline. With a capable hand on the shift lever, the five-speed transaxle clicks off clean shifts every time. A wonderfully composed chassis holds up its end of the bargain as well. As we reported last month, the 944 is a consummate handler, and its brakes are in perfect tune with the very capable suspension.

A few *C/D* pilots did complain about the way the 944 acknowledges every expansion joint with a loud smack and a sharp jolt, but all appreciated the well-controlled suspension strokes that enable the car to glide over larger bumps and dips. Some of us preferred the slightly more communicative manual steering to the lower effort and the faster response of the new power mecha-





PHOTOGRAPHY BY DICK KELLEY

nism, but eventually everyone accepted the new order. In the same way, initial misgivings about the tight thigh clearance between the steering-wheel rim and the seat gave way to eventual satisfaction with the race-car-like driving position. With their high, supportive side bolsters, the sport seats are a bit difficult to enter, but every member of the *C/D* test team—and we're talking about a wide variety of physiques—found a way to feel at home behind the wheel.

We accumulated miles rapidly at first, passing the 3000 mark by early April. Then tragedy struck. An elderly gentleman in a Chevette attempted to terminate his existence by pulling out of a driveway directly into the path of the 944. The skillful staff member at the wheel managed to avoid a deadly center punch, but the 944 did T-bone the Chevette in the rear fender. The perpetrator went to the hospital, his Chevette to the junkyard, and we lost five months while the 944's forward structure was rebuilt, at a cost of $6550.48.

The hard Michigan winter was approaching when we got the Porsche back in

**Vehicle type:** front-engine, rear-wheel-drive, 2+2-passenger, 3-door coupe
**Price as tested:** $23,155 (base price: $19,485)
Engine type: 4-in-line, aluminum block and head, Bosch Motronic engine-control system

| | | |
|---|---|---|
| Displacement | | 151 cu in, 2479cc |
| Power (SAE net) | | 143 bhp @ 5500 rpm |
| Transmission | | 5-speed |
| Wheelbase | | 94.5 in |
| Length | | 170.0 in |
| Curb weight | | 2820 lbs |

| **Performance:** | new | 30,000 miles |
|---|---|---|
| Zero to 60 mph | 7.4 sec | 7.5 sec |
| Zero to 100 mph | 21.7 sec | 22.4 sec |
| Standing ¼-mile | 15.5 sec | 15.5 sec |
| | @ 87 mph | @ 87 mph |
| Braking, 70–0 mph | 187 ft | 174 ft |
| Roadholding, 282-ft-dia skidpad | 0.83 g | 0.83 g |
| Top speed | 129 mph | NA |
| Road horsepower @ 50 mph | | 13.5 hp |
| EPA fuel economy, city driving | | 22 mpg |
| *C/D* observed fuel economy | | 21 mpg |
| Unscheduled oil additions | | 1 qt |

| **Service and repair stops:** | |
|---|---|
| Scheduled | 3 |
| Unscheduled | 5 |

| **Operating costs (for 30,000 miles):** | |
|---|---|
| Service | $352 |
| Repair | $1150 |
| Gasoline (@ $1.29 per gallon) | $1873 |

| **Life expectancies (extrapolated from 30,000-mile test):** | |
|---|---|
| Tires | 26,000 miles |
| Front brake pads | 30,000 miles |
| Rear brake pads | 92,000 miles |

early September, so we decided to try a new paint treatment to protect its finish. We chose TST Formula 5000 with Teflon, supplied by the Total Systems Technology dealer in Toledo, Ohio (800–245–4828); the manufacturer claims that this coating bonds to an automotive finish, keeping it bright and glossy for five years without any waxing or other maintenance. We don't know about the five years, but by mid-May 1984, nearly 30,000 miles later, most of the 944's finish looked as good as new. The only areas of the car that had suffered were the stone-chipped nose, where some of the repaired paint was flaking off, and the lower part of the front spoiler, which had seen one too many encounters with curbs. Since it's clear that Formula 5000 didn't hurt the

Porsche's finish, we intend to put it to a more definitive test, coating half of a long-term car with the treatment and maintaining the other half with conventional waxes.

Our experience confirmed the 944's promise as an even-tempered thoroughbred. We averaged 21 mpg—a touch below the 22-mpg EPA city rating, but an excellent figure in view of our frequent forays to the extremes of the car's performance envelope. Only once did we need to add a quart of oil between the 944's 15,000-mile oil changes. And most of the work at the rare scheduled services consisted only of minor checks and adjustments. Other than the lubricant, the only items calling for periodic replacement in the 944 are the spark plugs and the air filter (every 30,000 miles), the brake fluid (every two years), and the fuel filter and the exhaust-gas oxygen sensor (at 60,000 miles).

Of course, the life of certain components depends on the driving environment. Despite our repeated examinations of the Porsche's high adhesion limits, the original Pirelli P6 tires lasted for 26,000 miles—quite an acceptable life for a tire designed more for performance than for long wear. We replaced these with Goodyear Eagle GTs in the same 215/60R-15 size; they matched the Pirellis' skidpad adhesion but seemed a trifle softer in their on-center response. On the other hand, they softened the 944's tendency to hammer flat every expansion joint, and they also stuck tightly in the rain.

A retest after 30,000 miles showed that the Goodyears also improved the Porsche's braking. In addition, we discovered that the 944 had lost very little power over the long haul. Unfortunately, though, the engine's soothing low-rpm smoothness had vanished somewhere in the course of the 30,000 miles. We suspect that a loss of fluid in at least one of the hydraulic engine mounts was responsible.

A few other problems cropped up. At 12,000 miles, the power radio antenna and the driver's-side electric mirror failed; the mirror's wiring was repaired, and the antenna was replaced, both under warranty. At about 15,000 miles, the catch on the driver's door broke, a boot on the steering rack had torn and was leaking, and the clutch slave cylinder failed; again, all items were repaired under warranty. At 17,000 miles, the right-side windshield wiper came adrift; it was reinstalled and secured in place. At 30,000 miles, the power antenna was again on the blink, and the air conditioner needed recharging. Our 944 was also affected by a recall of all 1982–84 Porsches to have retractors installed on their rear seatbelts.

Although the 944 felt new-car solid throughout the test, we found its lengthy problem list a bit disconcerting. In its defense, only the clutch failure affected the car's mobility, and most repairs were covered by the warranty—which was fortunate

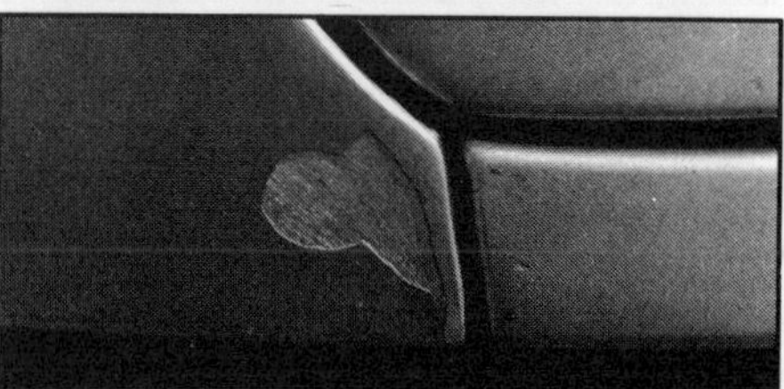

for us, because shop time at the Porsche dealer is never cheap. Each of the 15,000-mile services cost us about $170, mounting and balancing the four new tires cost $102, and wheel alignment (admittedly a front-and-rear project) cost $68.

This last operation was performed twice during the 944's life at *Car and Driver*, at 15,000 and 26,000 miles, because several staffers are very sensitive about a car's ability to track straight down the road. And all of us abhor the vibrations of imperfect wheels, which accounts for an unscheduled wheel balance at 14,000 miles.

Perhaps we were so picky because we all loved our 944 so much. Despite its problems, the logbook we kept in the glove box is loaded with superlatives. Certainly, if any of us were in the market for an affordable (current base price: $21,440, including power steering), low-maintenance, high-performance sports car, the 944 would be a strong contender. Evidently the rest of the world feels the same way: the waiting list at your neighborhood Porsche dealer is now longer than ever.

—*Csaba Csere*

# Callaway Turbo Porsches

*Rounding out Porsche's turbo lineup.*

*Twin turbos (within circles) and their associated hardware are squeezed under 928's hood.*

• Porsche may be the world's preeminent builder of turbocharged racing cars, but the firm's blower magic isn't always available in its street cars. In fact, there hasn't been a turbocharged factory Porsche in America since 1982, when the 924 Turbo gave way to the 944. If you crave a turbo Porsche, one solution is a gray-market 930 Turbo, but there are other ways to satisfy your appetite. Callaway Turbosystems of Lyme, Connecticut, offers aftermarket alternatives in the form of carefully engineered turbo kits for both the Porsche 944 and the Porsche 928. Since the 944 four-cylinder is essentially half of a 928 V-8, the systems are very similar. We sampled both on a visit to the Callaway works.

Surprisingly enough, very little mechanical refinement was lost in the turbo transformations. Both cars ran smoothly and quietly at low rpm, and we found no evidence of drivability glitches or detonation. When driven with a gentle foot, neither car betrayed its turbocharged nature. If the 944's tachometer was over 2500 rpm, however, flooring the throttle produced an irresistible forward thrust. The 928's turbo motor came on with a more frenzied action at a bit higher rpm. It seemed to hurl the car forward with its boost instead of simply pushing it ever faster.

Callaway does not know exactly how much power his blown engines develop, but, judging by their performance, it's substantially more than stock. As the accompanying chart shows, the turbo 944 is in a completely separate class from the standard car. In fact, it's substantially quicker than a standard 928S, and it will just about keep pace with a 911 Carrera. The turbo 928 is equally impressive, especially at high speeds, because its power never seems to run out. We had no chance to top out either car, but the turbo 928 was pulling harder at 130 mph than a BMW M1.

Such performance requires more than the simple installation of a turbocharger. On the 944, the 9.5:1 compression ratio is reduced. Callaway's technicians disassemble each 944 engine and trim the compression ratio to 8.1:1 by machining the combustion chambers and the pistons. A new exhaust system is fitted, complete with a manifold casting, an IHI RHB6 turbocharger with an integral waste gate, oversized exhaust pipes, and a low-restriction muffler. On the intake side, pressurized air flows through an intercooler in front of the radiator. (According to Callaway, his heat exchanger reduces the intake-air temperature by as much as 130 degrees at full boost.) The compressed and cooled air then flows through the 944's flow meter and intake manifold.

Callaway does not alter the Motronic engine-control computer's program, but he does fit higher-flow injectors and recalibrates the airflow meter to accommodate the increased volumetric needs of the blown motor. He also installs a microfueler, a computer-controlled fuel-injection system that adds an extra quantity of fuel during heavily boosted conditions.

Callaway uses two turbochargers for the 928. Unfortunately, the 928's densely packed engine compartment makes for a difficult installation. The only space available for the turbos is very low, near the back of the engine. Even this location is too cramped for integral waste gates, so two remote Callaway waste gates are fitted. The turbos are also too low for their oil to drain into the engine's oil pan, so a separate sump with its own scavenge pump is installed. The front-mounted intercooler and its plumbing are also a tight fit. Late-model 928s must undergo the compression-reduction machining; early models, with lower compression, can get by without it. The fuel-flow modifications are similar to the 944's except that a pair of microfuelers are fitted, one for each bank.

Such elaborate transformations don't come cheap. Callaway will charge $18,500 to turbocharge your 928, or $7850 to convert your 944. (If the machining isn't necessary on your 928, you'll save $2500.) In both cases, emissions certification is irretrievably lost, and some reliability is doubtless sacrificed as well. But the point is horsepower, which these blown engines deliver in prodigious amounts. So if you're tired of waiting for a factory turbo Porsche, the Reeves Callaway alternative may be worth considering.

—*Csaba Csere*

| Performance Comparison | horsepower (SAE net) | acceleration, sec | | |
|---|---|---|---|---|
| | | 0–60 mph | 0–100 mph | standing ¼-mile |
| **STOCK 944** | 143 | 7.4 | 21.7 | 15.5 @ 87 mph |
| **TURBO 944** | 215* | 5.5 | 13.9 | 14.2 @ 101 mph |
| **STOCK 928** | 234 | 6.2 | 17.8 | 14.7 @ 94 mph |
| **TURBO 928** | 300* | 5.6 | 13.3 | 14.1 @ 104 mph |

** C/D estimate*

# IMSA GTP Showdown, 1984

*The Kreepy Krauly March 83G and the LubeGard Porsche 962
face off with assorted nuts and sickies at the wheel.*

## BY LARRY GRIFFIN

• An amazing array of hardware has pitched tents in IMSA's opposing GTP encampments. Ford, Jaguar, Lola, March, Porsche, and a scattering of oddball hopefuls have settled in for a long war, and few of them appear short of ammunition. Three years ago we tested Brian Redman's Lola T600, which had locked up the IMSA title. Last year we tested Ford's Mustang GTP, which set *C/D*'s all-time low-speed-

skidpad record and promptly went out and won its first race. We vowed to test some more GTPs.

Late last spring, the phone rang. "This is LubeGard calling," said a well-oiled voice. Our receptionist supposed it might be an obscene call and put it through. No, Bruce Leven's Bayside Motorsports team was simply volunteering to truck its LubeGard Porsche 962 all the way from Redmond,

Washington, to our Ann Arbor offices.

At almost the same instant, another of our lines lit up. "This is Kreepy Krauly calling," said a disembodied voice. The receptionist shivered and put that call through, too. The Kreepy Krauly racing team was volunteering to truck its March 83G-4 all the way from Tucker, Georgia, for us.

The Porsche 962 is a close relative of the all-conquering 956, the best endurance

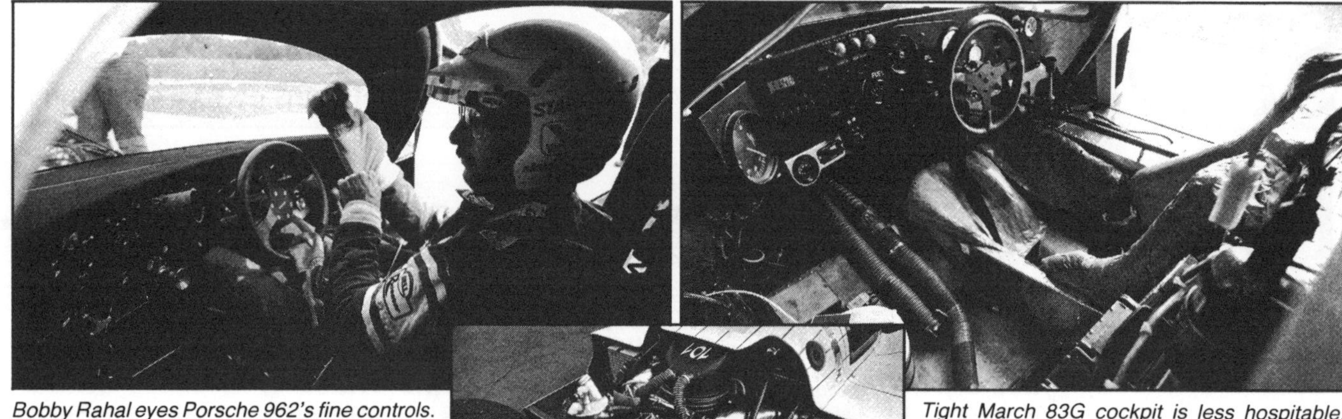

Bobby Rahal eyes Porsche 962's fine controls.

Long turbo runners of 962 add throttle lag.

Tight March 83G cockpit is less hospitable.

racing car ever fielded. In 1982, their first year at Le Mans, 956s finished one, two, three, all the more notable because they were the first ever monocoque-chassis, ground-effects cars from the Porsche factory. They have gone on to win virtually everything else in endurance racing, including Le Mans twice more, even though the quarter-million-dollar cars were in the hands of privateers for the 1984 race because Porsche had withdrawn its factory entry to protest the FIA's waffling fuel-consumption regulations.

Porsche was also irked by IMSA rules that prevented it from racing the 956 in America. IMSA has no limits on fuel, but it does have regulations that specify a car's minimum weight with various engine displacements and configurations. More important, IMSA insisted for safety's sake that the driver's feet not extend past the center line of the front axle. Porsche, of course, hoped IMSA's John Bishop would make it easy for the 956 to run. The German firm insisted that the 956 is safe despite pedals situated almost five inches forward of IMSA's limit, but IMSA showed no sympathy.

Desperate to race its great prototype in front of American audiences, Porsche was shown the way by its own Peter Schmitz. Since 1977, Schmitz has been the company's liaison with American Porsche teams.

"I realized by '82 that the future was for GTP cars," he says. "Our 935 was a pretty good car for its time, but I was looking around and seeing March-Chevrolets and Lola-Chevrolets and a March with a BMW engine. It was the European engine that actually woke up my first idea. I said, 'Why isn't it possible to run a turbo Porsche engine in a March?' I knew Porsche was busy with its 956 programs and the Formula 1 engine program, so I thought I might have a possibility to start a program that would support other teams that had the money to develop it. I talked to David Cowart, who had been running the BMW engine in his Red Lobster March, and he was interested. I measured the car and worked with Dave and with Alwin Springer at Andial in California to get the whole thing running. It

was hard, but I was gathering support and the idea was taking hold. Bayside came in with a Lola for our engine, and Al Holbert with his CRC March, and with more and more people interested, the factory could finally see a point in this. So, because of the IMSA rules and our engine work for others, the decision was made in October last year to build the 962."

The competition shuddered. Porsche teams rejoiced. The Porsche single-turbo, single-overhead-cam, 930-based engine could be grafted into the basic 956 package that had utterly proved itself. But time was short. Construction of the first 962 started late in November, and the car was scheduled to run at the Daytona 24-Hour in early February.

"To begin with," wrote chief 956/962 architect Norbert Singer in *Christophorus* magazine, "the front axle had to be pushed forward by 12cm. This gave 12cm more wheelbase, with a new, longer monocoque. Since overall length had to remain the same, front overhang was shortened. This had an influence on the aerodynamics of both bottom and top surfaces. The longer wheelbase also required changes to the center portion of the car. The tail had to be altered as well, to fit IMSA's rule book, since aerodynamic aids—including rear wings and fins—cannot extend beyond the vehicle silhouette [as seen from above]."

Beneath the 962's Kevlar skin is Porsche's second ever monocoque chassis design, its 110.0-inch wheelbase fitted with such top-flight running gear as front and rear unequal-length control arms, titanium coil springs, and anti-sway bars. The shock absorbers are gas-pressurized Bilsteins. The brakes are 13.0-by-1.2-inch vented discs with two calipers per rotor both front and rear. The BBS wheels are 13.5- and

14.0-by-16s front and rear, and the tires are Goodyears. They're guided by rack-and-pinion steering that provides only 1.5 turns lock-to-lock. The bell housing, the all-synchro gearbox, and the basic suspension design were carried over from the 956 (though new spring rates and shock valvings were needed since the 962 ground-effects tunnels and bodywork provide less downforce).

If Porsche were to use a 956 engine, with dual plugs, four valves per cylinder, water-cooled heads, and twin turbos for the 962, it would have to shrink from a ripe 2.7 liters to a wimpy 2.1 liters to race in IMSA. Porsche finally settled on a 2.9-liter version of its flat six, with two valves per cylinder and the single turbo. After early development, the 962 boasted 651 horsepower at 8000 rpm and weighed 110 pounds less than would an IMSA-legal 962 with the smaller-displacement twin-turbo engine.

"The 930 Turbo street engine is the base," says Peter Schmitz. "But we never had single-plug-head experience from the days of racing 935s. We were fortunate to have the 956 electronics, or it could have been very expensive. We finally developed a reliable, no-popping engine. The electronic programming eliminates a big part of this, and we have a new intake manifold and pistons and a different combustion chamber for better swirl. One electronic chip controls the injection system, the ignition, and about eight other functions, and with it we can check the whole engine in the pits. A memory function recalls rpm, intake temp, ignition timing, boost pressure, injection timing, and a few other things."

Thanks to a lot of midnight oil, the 962 was ready for Daytona. Unfortunately, Porsche infuriated the faithful American teams that had long ago placed 962 orders: only one car was built, and it was in the factory's hands. With no time for serious development, Porsche plugged Mario and Michael Andretti into the cockpit and hoped for the best. Mario put the car on the pole by almost two seconds, and he and his son led the early stages of the 24-hour race. But to the glee of the left-behind, the 962 suppered early, swallowing a few gearbox bits as an appetizer and then finishing off its nonpopping engine with a major belch.

Although the 962 had failed its first big test, it had also shown its appetite for running up front. After Daytona, the privateers began to get their long-awaited 962s —Bruce Leven, Bob Akin, Al Holbert, and more on order. There was a sensational six-hour race at Riverside between Leven's car, driven by Holbert and Derek Bell, and the March-Chevy of Bill Whittington and Randy Lanier. The race was decided at its very end, when a screw-up with the Leven team's radio brought an unannounced last-minute pit stop. Whittington and Lanier won by six seconds, but only because they'd made their own luck. With his own 962 due for delivery soon, Holbert took his Löwenbräu-beer sponsorship money and left Leven to fend for himself. Leven still had his sponsorship money from Lube-Gard, which manufactures a biosynthetic-lubricant additive that the team has been using in its engines for several years, and some of that money quickly went into hiring Indy ace Bobby Rahal and enduro prince Hurley Haywood.

It was Haywood the Florida golden boy, former winner of Daytona and Le Mans and countless other GT contests, who showed up for our 962 test. After a monstrous crash at Sears Point, he'd been on a crutch for almost a year. Haywood was well tanned, but his lower leg was full of rods, screws, a plate, and two drainage holes, its Ace-wrapped scars graying around the edges like spoiling tuna. The doctors had him on low-grade antibiotics four times a day, but his powers of observation were

*Hurley Haywood's leg still hurts (note duck).*

fine as he talked about his Le Mans–winning 956 versus the March.

"It's kind of like a Cadillac versus a truck in ride characteristics. In the Porsche, everything fits beautifully, all the switches are right where you want them, the seat is very comfortable, the steering is very easy, the pedals are in the right position. The March is more of a raw car. Its seat sits down be-

*South Africa's van der Merwe loves triumphs.*

tween two bulkheads at an angle to the wheel. The Porsche might corner a little better through the high-speed stuff, but a Chevy engine in the March makes it an easy, fun car to drive.

"The March is also very reliable and easy to work on. There's no big mystery. When you take off the lid and look at a 962, you're just flabbergasted by all the pipes and all the complexities of the motor. For a March-Chevy, you're talking about an initial outlay below $150,000, whereas the Porsche is going to be $250,000 just for starters. But Porsche is hard to beat in the long run.

"At Le Mans, the 956 was enough to make you gasp for air," said Haywood. "I've never driven a race car that had that kind of cornering ability. If you're talking about just the chassis, I would say the Porsche is much better than the March. The Porsche just feels like it's glued onto the track, whereas the March can be very tipsy. In braking, the two are probably equal, but I've mashed as hard as I can on the March brakes, and the thing just wouldn't come to a stop. Porsche brakes feel like power brakes practically, and all-around comfort is like a passenger car. Little things that may sound stupid are important, and simply being comfortable is worth a half-second per lap."

Unfortunately, even Porsche's ergonomics could not help Haywood's ailing leg. He was in much greater pain than he had let on to Leven, and to everyone's worried disbelief he was visibly slow around the small skidpad and in the acceleration runs. He came to life briefly on the high-speed skidpad, running hard but stopping to sigh, "Well, if I just knew how fast the others that you've tested ran here . . ."

That wasn't the point, but then neither was it to poke a hot stick into his wounds. We dropped Haywood at the airport, and later, when all the results had been tabulated, let Leven and Schmitz know that the car almost certainly had more health in it than a sick man could draw out of it. In a few weeks, they would ante up a very healthy driver, Bobby Rahal.

In the meantime, we trundled off to the Transportation Research Center in Ohio to meet the March-Porsche. After qualify-

ing second to the Andrettis' 962 at Daytona, it had won the race, and you might say its drivers had come out of the blue. The car came from South Africa, and it was called "the Kreepy Krauly" by the crowd, but it wasn't a newcomer: last year it had carried Al Holbert to the IMSA championship. Now it was colored aqua, royal blue, and white, like water shimmering in a swimming pool—good graphics for a car sponsored by a South African–based manufacturer and distributor of automatic swimming-pool cleaners called Kreepy Kraulies. Kreepy Krauly is flooding the American market even now.

Like the Porsche 962, the Kreepy Krauly March has a single-turbo Porsche engine, albeit a bigger 3.2-liter unit, whooshing with an added 69 horsepower for a total of 720 at 7000 rpm. The March's owner and primary driver is Sarel van der Merwe, and he obviously likes nothing better than all that fun. Since Jody Scheckter's retirement, van der Merwe, an eight-time national rally champion (seven in succession) in everything from Datsuns and BDA Escorts to Audi Quattros, has become South Africa's leading racing driver, as the GTP crowd found in a fearsome flash.

Six feet two inches, lean, dark, sharp-featured, quick-eyed, crisply mustached, van der Merwe looks as if he should be wearing tailored khakis, rows of medals, and a saucer cap with a spit-shined brim. But his eyes would give him away—fierceness layered over and touched with humor. He once got a $1000 speeding fine.

Once upon a time early in the season, van der Merwe had a March 84G. It burned at Road Atlanta, so the crew updated the old 83G he had shared with Graham Duxbury and Tony Martin to win Daytona. The week before we tested, van der Merwe and the 83G won again, this time in the downpour at Lime Rock. The raunchy rally driver in van der Merwe's sideways soul loves the beat of steady rain as much as most road racers hate it.

## Racing around the Corner

**OCTOBER**

- **14** CART/PPG Miller High Life 150, Phoenix International Raceway, Phoenix, Ariz.
- **19** NHRA Winston World Finals, Pomona, Calif.
- **21** CART/PPG Laguna Seca 300, Laguna Seca Raceway, Monterey, Calif.
- **21** NASCAR Warner W. Hodgdon American 500, North Carolina Motor Speedway, Rockingham, N.C.
- **26** SCCA Press on Regardless Rally, Houghton, Mich.

**NOVEMBER**

- **1–4** SCORE Baja 1000, Baja, Mexico
- **4** NASCAR Atlanta Journal 500, Atlanta International Raceway, Atlanta, Ga.
- **10** CART/PPG Caesars Palace Grand Prix, Caesars Palace, Las Vegas, Nev.
- **17** SCCA Oregon Trail Rally, Tualatin, Oreg.
- **18** NASCAR Winston Western 500, Riverside International Raceway, Riverside, Calif.
- **25** IMSA Camel GT, Champion Spark Plug Challenge, Kelly American Challenge, Daytona International Speedway, Daytona Beach, Fla.
- **25** FIA World Rally, Chester, England
- **30** SCORE Barstow Classic, Barstow, Calif.

# Porsche Potency

*Belts! Chains! Four cams! 32 valves! 288 horsepower!*

BY DON SHERMAN

• The tide has turned. Never before in recorded history has a European car manufacturer tooled up such an elaborate package of technical advances and aimed it first and foremost at America. For the next two years or so, Porsche will be shipping 928 models to the U.S. with sophisticated twin-cam, four-valve cylinder heads *that will not be available in its home market.*

This radical move is a creative way to pump a bit more life into Porsche's flagship and simultaneously to throw the gray market a surprise curve ball. Thanks to the new under-hood technology, the 928's horsepower, torque, fuel economy, acceleration, and top speed are all significantly improved. Naturally, the price will take a fat hike as well: the final figures haven't yet been released, but John Cook, president of Porsche Cars North America, advises that the window stickers for the 1985 models will be pushing $50,000, an increase of roughly $5000.

In joining the rapidly growing four-valve-per-cylinder club (Ferrari, Lotus, Saab, and Toyota have four-valve engines on the U.S. market, and Jaguar, Mercedes, Oldsmobile, and others will join them in the near future), Porsche has added a few of its own technical twists to the 79-year-old idea. Though the 928's pent-roof combustion chambers, centralized spark-plug location, notched pistons, and free-flowing ports are standard practice with four-valve designs, the visionaries of Weissach have come up with what we believe is a unique system of driving the quartet of camshafts.

The new cam-drive system was crucial to Porsche's goal of preserving as much of the existing all-alloy V-8 as possible. Obviously, new cylinder-head castings were a must, but Porsche engineers, led by Paul Hensler, the firm's director of powertrain design and development, did manage to salvage nearly every last detail of the original engine's toothed-rubber-belt drive mechanism. This accomplishment was significant because a new layout in this area would have scrambled all the accessory drives and demanded a new design for the front half of the engine.

Several fresh features have been added to the layout, however. The drive belt has stronger reinforcement fibers, and its tension is now automatically maintained by a hydraulically damped bimetallic-strip device. (Similar materials are used in home thermostats; in the Porsche's case, heat causes the tensioner to shift position and take up any slack in the cam belt.) The new exhaust camshafts are driven directly by the long rubber belt, and they lie in the same location used in the single-overhead-cam head. The intake camshafts are positioned several inches inboard and are driven by a short run of single-row roller chain that loops around both of the camshafts in each head (see drawing). Each of the chains is positioned centrally along the length of its head and has its own automatic tensioner-damper device.

Hydraulic lifters between the cam lobes and the valves make periodic lash adjustment unnecessary. Many manufacturers favor a separate, bolt-on carrier for the camshafts and the lifters, but Porsche has incorporated this function in the elaborate aluminum cylinder-head castings.

Unfortunately, you can't see much of the heads' mechanical beauty when you raise the hood: they are shrouded by an intake manifold that looks like something designed to accompany the Mormon Tabernacle Choir. It comprises two massive plenums running lengthwise, a huge air-filter housing across the back, eight pipe-organ-like runners, and one more tube that connects the plenums with a central air meter. To take maximum advantage of the harmonic-resonance effects inside the intake manifold, no two consecutively firing cylinders draw from the same plenum. This system, together with the 928's low hood, necessitates two different tuned lengths, a design feature that in itself helps spread the ram-tuning torque boost over a broader rpm range. To save weight, all the intake-manifold pieces are cast magnesium.

The new race-bred valvetrain is only one item in a long list of updates for the 928's V-8. The cylinder bores are three millimeters larger, bumping the displacement from 4.7 to 5.0 liters. The main-bearing webs are meatier so that they can withstand higher loadings. The compression ratio has been increased to 10.0:1 (up from last year's 9.3:1). The Bosch electronic fuel injection is now signaled by a hot-wire mass-airflow sensor. The ignition is the latest Bosch Motronic system; it selects from a map of 256 combinations of spark advance and fuel-air ratio. The new exhaust system is a true dual-pipe design, featuring lightweight stainless-steel-tubing headers, a catalyst that is much larger in frontal area and volume, a higher loading of platinum and rhodium (the noble metals inside a catalytic converter), and a twelve-percent reduction in full-load back pressure.

The fancy hardware does indeed make a significant difference, both in what goes into the engine and in what comes out. Fuel economy is slightly better with either the five-speed manual or the four-speed automatic (although the correction factors used to determine the 1985 EPA ratings produce lower window-sticker figures). The torque curve now has a nice, fat hump at a very usable 2800 rpm, and it's both flatter (overall) and higher on the scale than

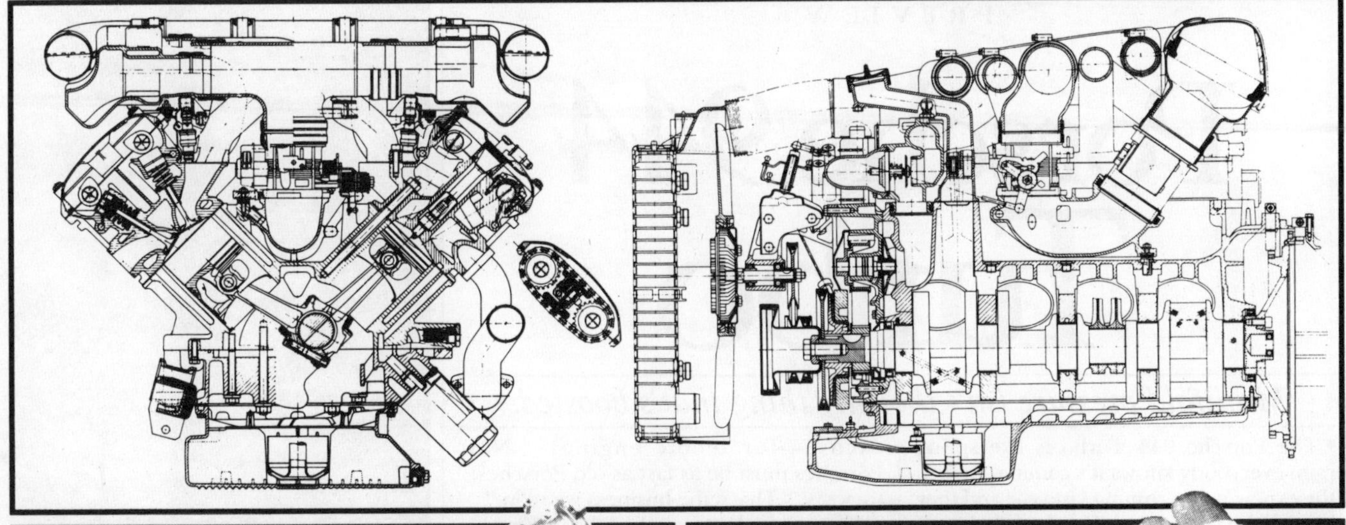

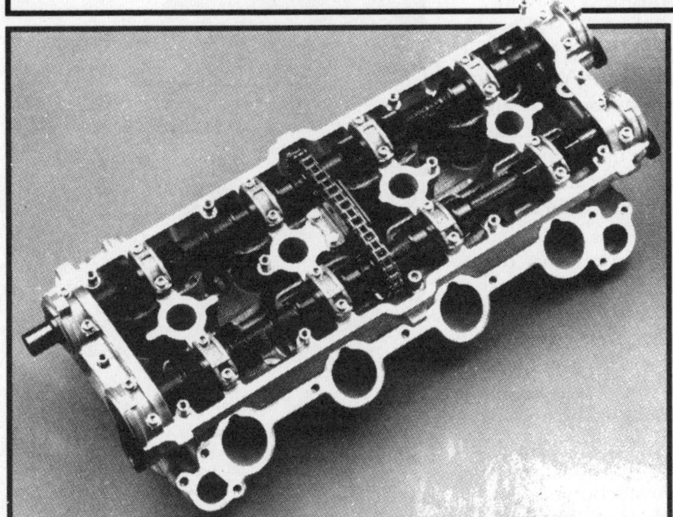

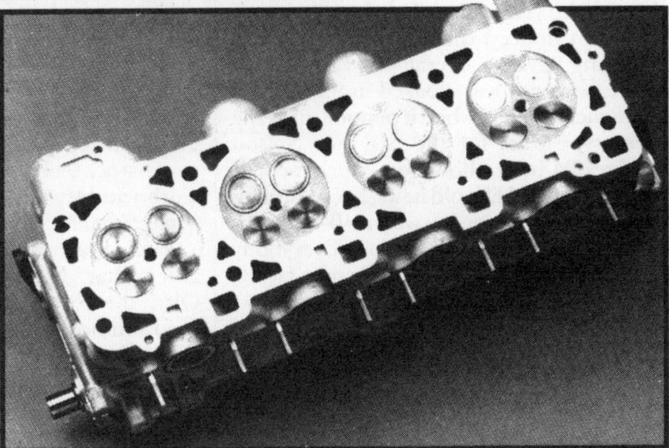

Porsche's clever two-stage cam-drive system allowed the addition of twin-cam cylinder heads without a complete redesign of the 928 V-8.

the 4.7-liter engine's output curve. Rated horsepower is up to 288 hp, a 23-percent improvement, and its peaking speed has been raised by 500 rpm.

Although we haven't yet been given an opportunity to conduct full test procedures, our initial driving experiences in American-spec 928s and Porsche's own performance figures make it clear that this car is now a proud member of the 150-mph club. The manual-transmission version is particularly brilliant in its throttle response, whether you're chugging along at 2000 rpm in third or hammering at the atmosphere at 4500 rpm in fifth.

The rationale for the new four-valve engine goes well beyond simple torque-curve plumping. A very potent 944 Turbo will be here in a few months—capable of exceeding 150 mph, according to reliable reports—and it wouldn't be appropriate for that upstart to fly by Porsche's flagship in top-speed and acceleration performance. Furthermore, it is Porsche's stated intention to eliminate systematically the performance differences that currently exist between U.S.-spec and European models. (The new Turbo will be a "world" car, equipped with the same basic hardware for all markets.) Equalizing U.S. and German

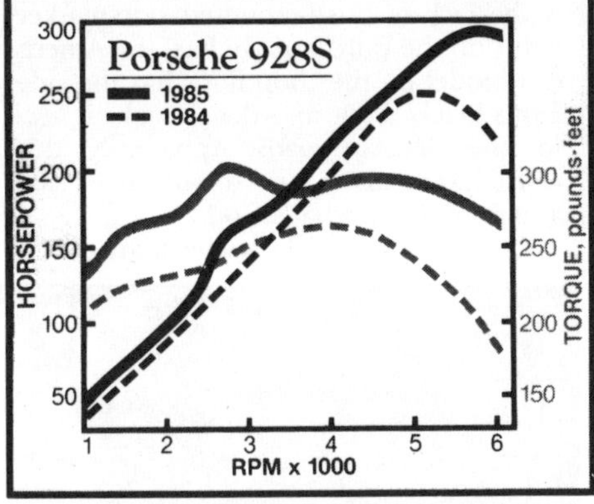


| Engine type | V-8, aluminum block and heads |
|---|---|
| Bore x stroke | 3.94 x 3.11 in, 100.0 x 78.9mm |
| Displacement | 303 cu in, 4957cc |
| Compression ratio | 10.0:1 |
| Engine-control system | Bosch Motronic |
| Valve gear | belt-and-chain-driven double overhead cams, hydraulic lifters, four valves per cylinder |
| Power (SAE net) | 288 bhp @ 5750 rpm |
| Torque (SAE net) | 302 lb-ft @ 2400 rpm |
| Redline | 6000 rpm |

| Mfr.'s performance ratings: | auto | manual |
|---|---|---|
| Zero to 60 mph | 6.6 sec | 6.1 sec |
| Standing ¼-mile | 14.9 sec | 14.2 sec |
| Top-gear passing, 30–50 mph | 2.2 sec | 8.2 sec |
| 50–70 mph | 3.0 sec | 8.6 sec |
| Top speed | 152 mph | 155 mph |
| Fuel economy, EPA city | 16 mpg | 15 mpg |
| EPA highway | 22 mpg | 22 mpg |

performance was no small feat in the 928's case, since it has been such a strong autobahn runner for so long. Fortunately, instead of slowing the home-market car down to U.S. levels of speediness, Porsche undertook the more ambitious task of pulling the American model up by its bootstraps. This upgrading is also a shrewd investment in the future, because the new four-valve head will fit the 944 engine as well; we will likely see a sixteen-valve four-cylinder from Porsche within two years. By then, the four-valve crowd will be much larger. Today, however, racy cylinder heads are still very special, and they do go a long way toward making $50,000 for an automobile seem almost reasonable.

Unfortunately, there remains one small chink in Porsche's armor. The European 928S is still a superior automobile because anti-lock brakes were added to its standard-equipment list for the 1984 model year. We feel strongly that such hardware should be part of the 288-horsepower, 155-mph, $50,000 deal in the U.S., and that opinion has been registered with Mr. Cook of PCNA. Porsche's best customers are here in America, and they deserve the best 928s that the Zuffenhausen factory is capable of building. ●

# Porsche 944 Turbo

## *Manifold pressure isn't the only thing that's boosted.*

• The Porsche 944 Turbo is like spring rain: everybody knows it's coming. In fact, they knew it was coming last year and they knew it the year before that. Astute observers even noticed the engine racing—and finishing seventh overall—at Le Mans in 1981, where it was cloaked in 924 Turbo nomenclature. So when it popped up in the Prototype class for the past two Nelson Ledges 24-hour Showroom Stock races, it kind of seemed like old news. Even winning the event last summer, with a 40-lap cushion, didn't cause much ruckus. Everybody knows Porsches are supposed to be fast.

So the 944 Turbo now makes its showroom debut amid shrugs of anticlimax. Until you drive it. Then it's *holy tire smoke! This little mutha hauls zass!*

Never mentioned in all the prelim-speak was that the 944 Turbo would have a 911 Carrera level of horsepower: 217 SAE net, to be exact. And the 911 goes like a scalded dog. We found out about the Turbo's bounty at the Frankfurt show late in 1983 in a casual conversation with Paul Hensler, Porsche's director of powertrain development. Can you imagine the scalpel-sharp handling of the 944 with all that horsepower? We could, and it was more than patience could bear. So about the time we thought pre-production samples would be plentiful in Weissach, we announced to Porsche that we were coming, cameras and hot corpuscles at the ready.

We arrived, and so did the biggest European snowstorm in at least a generation. At Weissach, the shop mechanics were fitting chains onto a 911. Honest. So much for the hot laps. But at least Paul Hensler and Helmuth Bott, chief of R&D, and Hans Mezger, most recently the father of the TAG Formula 1 engine, were on hand to answer questions.

The 944 Turbo is an engineers' car, and Porsche engineers are pleased that it's the first model they know of, from their shop or anybody else's, where the catalyst version and the noncatalyst version have the same power. Bott and Hensler speak with pride on this point. All Germany is now in an emissions-control mood, moving to catalysts and unleaded fuel. But Porsche engineers don't see that as a reason—or an

excuse—for feeble engines. "New Porsches must be as fast as old Porsches," they say. "That's the business we're in."

Those who have watched the 911 evolve over the past twenty years know it's the engineers, not the stylists, who call the cadence of change at Porsche. So the changes tend to be functional. And subtle. It's a great self-test to conjure up an image of a stone-stock 944, then wander around a 944 Turbo and try to figure out, *What's changed in this picture?*

Well, the nose is different. No doubt about that. No bumper on the Turbo. Okay, there's a bumper, but it's hidden behind a smooth, one-piece, flexi-plastic nose panel that makes it look as if somebody peeled off the bumper and filled in the attachment holes. Incidentally, the European and American versions are identical in this area; the engineers say the European customers like the extra protection of the U.S. standards. (In back, the U.S. bumper extends farther.) The only difference in front is the lack of flush-mounted side-marker lights on the Euro cars. In fact, the American model on the shop floor has the side lights blacked out in order to make it *legal* for the German roads. Apparently, this world has as many ideas about what's safe as it does about what's God.

Of course, the Turbo's wheels are differ-

  **PHOTOGRAPHY BY MARTYN GODDARD**

 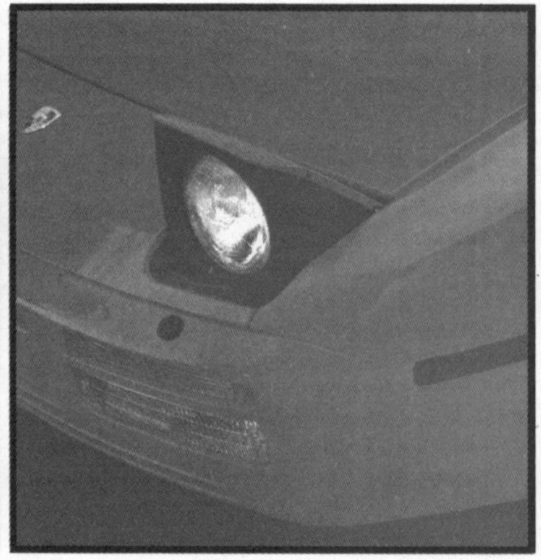 

ent from the 944's: sixteen-inchers with five bean-shaped holes, similar to a style introduced on the 928. The widths are seven inches front, eight in the rear. The tires are 205/55VR-16 Pirelli P7s in front, 225/50VR-16 behind. If you're really paying attention, you notice, just peeking through the bean holes, a sexy black caliper with raised letters machined off to say "Porsche" in contrasting bare aluminum; the engineers confirm that the brakes are new. *Auf Wiedersehen,* sliding calipers. *Guten Tag,* light-alloy, four-piston grabbers. The leading pistons are even different in diameter from the trailing ones to minimize ramp wear in the pads. Neat!

Porsche is a longtime believer in aerodynamics, so you expect refinement on this frontier as well. But the Turbo doesn't look much different from the regular 944. The nose is smoother, which should help, but compared with the whale-tail innovations of the past, the Turbo offers little to pin your hopes on. The only thing really conspicuous is the spoiler beneath the rear bumper. Certainly, this area has been a styling problem on the 944. Following along behind, it looks as if the manufacturing plant just ran out of bodywork when it got down to finishing the rear bumper. You see all kinds of undercoated stuff hanging down: suspension bits, a muffler, a fuel tank, and other unmentionables. On German roads, you see 944s fitted with various sorts of under-panels in an attempt to clothe the unsightly. The Turbo's solution is a conspicuously added-on panel that shields the view with about the same elegance that a barrel dresses up a nudist. But

# Porsche 959

## *Kicking the tires on the Mystery Ship.*

• In toting up the secrets that Porsche is not too good at keeping these days, the 959 has to go right to the top of the list. This is the model that was known as Gruppe B in its show-car days of eighteen months ago, a softly flared 911 with 400 hp worth of twin-turbocharged flat six in back, featherweight Kevlar bodywork on top, and electronically controlled four-wheel drive underneath to make it stick to the pavement like a bad reputation. Normally, whether you're talking car business or military, projects like this would be labeled "Top Secret" right up until they were operational. The only exception might be if you thought the project was so far beyond the enemy that he wouldn't be capable of a countermove—even if he had a blueprint. The 959 sounds like that sort of quantum leap.

Since Porsche is perfectly willing to talk, we made special efforts while we were in Weissach to listen. Helmuth Bott says that the 959 is a research vehicle meant to lead the way in what amounts to a technical trickle-down campaign for Porsche models for the rest of this century. But it's specifically intended for the FIA's Group B class, which allows any automobile, no matter how sophisticated, to participate in international races and rallies once 200 copies of it have been built. The 959 is *exactly* a 200-car program. Bott does leave the door open for twenty more cars, again for FIA reasons: once the 200 requirement has been met, the rules allow an even more potent version to compete as an "evolution of the type" once twenty examples of it have been produced. But the evolution card can be played only once, so 220 cars is the most 959s that will ever be built. We get the idea that Bott thinks the engineering department will have an even better idea after the production of the first 200 B-cars gets under way.

Since 200 is more racers than the world has any use for, the bulk of that number will be sold for street purposes. And not as hair-shirt transportation, either: automatic temperature control and a leather interior will be optional. You expect such niceties when a car stickers in at executive-jet prices: a base 959 is now on the quote board at 398,000 deutsche marks, which

There are more aero advancements, but they don't scream at you. A plastic tray fits under the nose, sealing tightly against the bumper skirt at its forward edge so that air is directed smoothly under the car. Just forward of the rear wheel, another small fairing at the bottom of the fender opening directs air around the tire. And the windshield is said to be "flush-mounted," though there is still a molding on the outside and a noticeable step between the glass and the sheetmetal. Taken altogether, these changes drop the drag coefficient from 0.35 to 0.33—not a big deal when confronting you from a magazine page but still a step in the right direction. With the fat P7s replacing 60-series rubber, the drag coefficient could very easily have gotten worse, had all this effort not been expended. Porsche engineers are right when they say it's very difficult to make a short car like the 944 score well in the wind tunnel. They go on to claim that the net drag of the 944 is lower than that of any of the sedans that advertise better coefficients. Drag is the product of the coefficient and the frontal area, and the latter is a mere 20.3 square feet on the 944.

Not all of the airflow-management efforts have been directed at drag reduction, however. A duct behind each end of the lower part of the grille directs cooling air

the engineers are happy with the result. They say it speeds up the exit of air from under the car, which cuts drag; it enhances crosswind stability; and it improves the ventilation of the muffler–gas-tank–differential area, lowering the operating temperatures of those components. We would say that it also attracts the eye, much as spoilers on top of the car do, and could well turn into the fashion accessory of the late Eighties, never mind that it fits like a shoe on a turkey.

is $127,000 at today's exchange rate.

Don't change your money just yet, however. When production starts in September, there won't be any U.S.-spec models. The car is being developed with catalysts to handle expected German emissions standards—and will presumably not be far off U.S. emissions requirements as it leaves the factory door. But no work is being done toward American bumper or crash requirements. Bott says that such development would take an extra year and is just out of the question.

The 959 is basically a 911. In fact, some of the prototypes around the workshop are obvious conversions. But the real thing has only a skeletal resemblance. The skin is a thin molding of Kevlar-reinforced plastic, which attaches to a modified 911 underbody. Of the exterior panels, only the doors and the front deck are metal, and they are aluminum.

Some of the differences represent major changes in 911 tradition. No more MacPherson struts in front and trailing arms in back: a pair of control arms now take up station at each corner of the car. Coil-over shocks are used; to keep them short enough so they don't poke through the hood in front, the work is divided among two units per wheel. The tires are Dunlop Denloc D4s, designed to maintain a high degree of handling even when deflated. Their sizes are 235/40 front, 255/40 rear, both VR-rated seventeen-inchers.

The engine has electronic controls for fuel, timing, and boost, patterned after those which give excellent track fuel economy to the 956 racer. The hole in the leading edge of each rear fender admits air to the intercooler in each rear corner. The twin turbos are located at the rear of the engine, just inboard of the coolers.

Still, it's the 959's drivetrain that really breaks new ground. The transmission is a manual six-speed, done purely to gain one more ratio. With top speed pushed up into the 190-mph range, Bott says, five speeds simply aren't enough to cover the territory.

Like the 911, the 959 has its engine in the rear, with the differential in front and the transmission in front of that. For four-wheel drive, a prop shaft pokes forward out of the transmission to drive a front differential. The shaft is contained in a rigid torque tube that bolts directly to the front and rear housings, making the whole drivetrain a one-piece affair. So far, this is about what you'd expect of a four-wheel-drive Porsche.

The trick comes in the torque biasing: to make such a car really handle, some road surfaces would like more power to the front wheels than others. Here is the point where the 959 truly operates on the fringes of technology. Porsche's scheme for torque biasing centers on a multi-disc clutch pack controlled by electronics, in effect a limited-slip differential with electronics deciding how much slip you should have at any given moment. On the 959, one of these units is between the rear wheels,

and one is in the shaft going to the front. Yet another between the front wheels has been used with success on the Paris-Dakar Rally cars and may be used on the 959 evolution. Meanwhile, two of these smart diffs provide more than enough variables for the engineering department to chew on.

The problem with electronics is that they are dumb. Somebody has to tell them what to do. Bott says they'll need separate torque-biasing programs for dry pavement, wet pavement, snow, gravel, who knows what: probably at least four different conditions. The car will have an automatic mode and then one where the driver can manually select among the different programs. The only problem is that engineering is a little behind in coming up with the programs. This has to be done experimentally, establishing the points on the map: try something, drive it, then try something else. Work fell way behind during the metalworkers' strike last summer, and now the world is covered in Cool Whip. At least the snow program can be worked out now, but some of the other programs will have to wait for the spring thaw. Which is going to make the September production target a nip-and-tuck deal.

Still, even if the worst case happens, Bott doesn't sound as if he'll miss his deadline by more than a couple weeks. He obviously knows the dimensions of the problem; in fact, they may even be what gives him the confidence to talk. He knows the enemy can't possibly catch up.  —*PB*

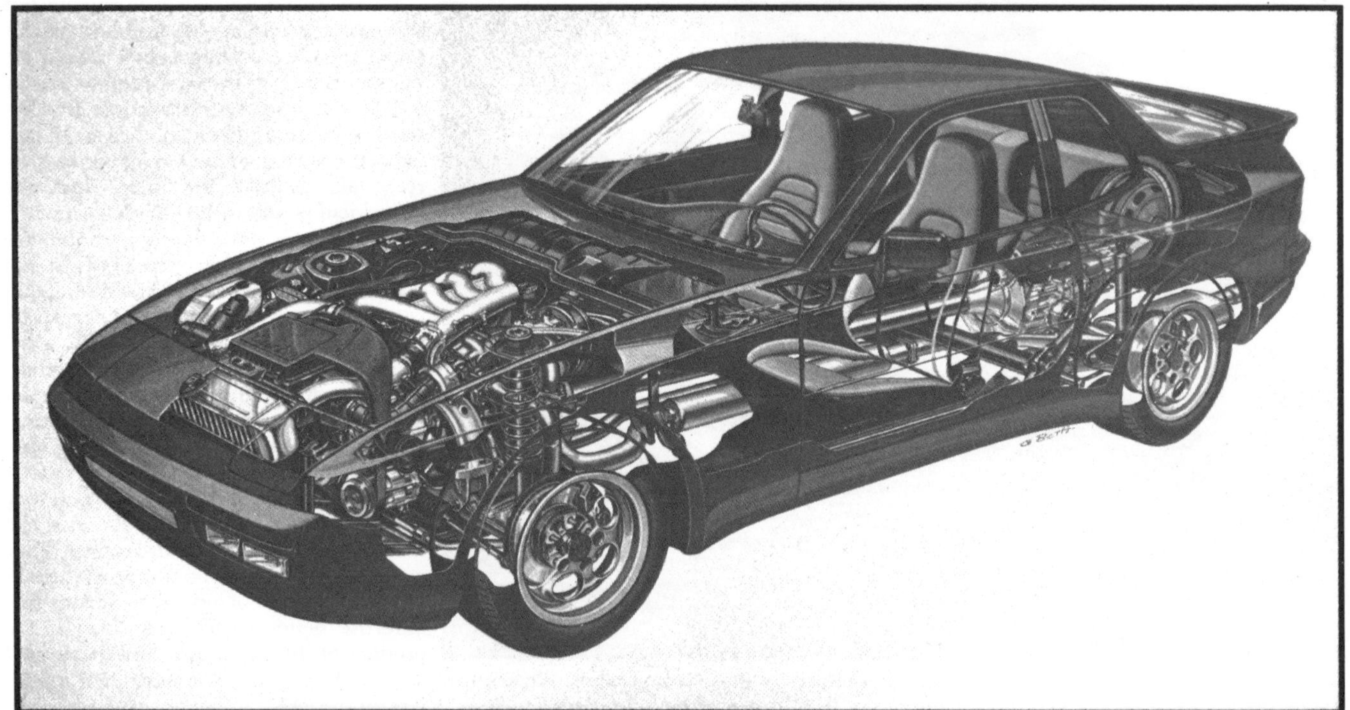

toward the front brakes. There is no actual connection of the ducts to the brakes, as there would be in a race car, but a baffle at the bottom of each MacPherson strut funnels the stream toward the rotor.

The big news in the cockpit is the redesigned dash—still very much in the Porsche tradition, we're pleased to say, which means no green-glowing digits to remind you of a video arcade and no voice synthesizers to remind you of technology running amok. No yellow markings on the instruments as in the current 944, either, and we're glad of that. Just white on black now, in four dials neatly arrayed before you, and huge vents that promise to move more air at less velocity and therefore with less rushing sound. The dashboard molding itself is of a somewhat different shape but no different effect. The console has been cleaned up noticeably, however, and it integrates better with its surroundings.

But you can't see the one change that Porsche engineers seem proudest of: the steering column has been raised 18mm (0.7 inch). The 924/944 has always had a semi-wonderful wheel position. Wonderful because the column is more horizontal than the columns of most cars, which means that the wheel itself is more vertical, so you don't have to stretch for the top of the rim when you're already strung out on the edge of control. But the horizontal col-

# The 944 Turbo Engine

*Wherein Porsche pushes four-cylinder horsepower through the 200 barrier.*

• Porsche can legitimately lay claim to more turbocharger experience than any other manufacturer in the world. Its 911 Turbo kicked off the modern era of turbocharging ten years ago, it was the first maker to apply intercooling to a production automobile, and its turbo engines have dominated racetracks all over the globe for thirteen years. Not surprisingly, Porsche management decided some time ago to design no new engine without considering the ramifications of a turbocharged version. And now that the 944 has seen three and a half years of production in normally aspirated trim, the time to build some boost under its hood has finally arrived.

Since the engineers knew what was coming when the 944's all-aluminum, single-overhead-cam, four-cylinder engine was derived from the 928 V-8 in 1981, it wasn't necessary to redesign any major components to support the addition of a turbocharger. Instead, Porsche's efforts were focused on adding new intake and exhaust

umn has its negative side, too: the bottom of the wheel is low, which means that drivers thick of thigh have no leg clearance. Porsche tried to alleviate the bind by offsetting the center of the steering wheel 20mm upward from the center of the column, which was fine as long as nobody tried to park. Fixing the rub required a new dashboard. Then, once the wheel was up, the seat needed an up-and-down adjuster (preferably power in this price class) so that short people wouldn't have to sit so low that they couldn't see. All of this has been accomplished in the Turbo, and the improvement is immediately noticeable.

At the start we mentioned the snow—fair warning that most of the Turbo's other details might not be noticeable under the circumstances. Gas-pressure shocks, for example: until we can see the ground again, we'll just have to take the engineers' word for their goodness. We can report, however, that the engine would bust the rear wheels loose in third gear. Probably it wouldn't have taken the Turbo's 78-percent torque increase over the regular 944 to produce that result, but nobody buys a Porsche just to get by. This is one seriously fast car, perfectly capable of melting its own snow so that it can prove the point.

Production was scheduled to start in February, slowly at first in order to maintain quality with all the new hardware. Very soon the factory will have the capacity to produce 150 944s a day. But rather than lock itself into a certain proportion of Turbos by advance decree, Porsche has decided to wait to see what the market says.

It's easy to say that this is a heck of a car and assume it will sell like mums on Mother's Day. But there's one hitch. Along with pumping the horsepower to the 911 level, Porsche is similarly jacking the price. It will be lower in the U.S., maybe a grand lower, but still over 30 thou. We may be flying blind in ever deepening snow, but we can see that such a price tends to freeze out the ordinary thrill seeker. —*Patrick Bedard*

---

**Vehicle type:** front-engine, rear-wheel-drive, 2 + 2-passenger, 3-door coupe

**Estimated price:** $31,000

**Standard features:** air conditioning with automatic temperature control, 6-way power driver's seat, electric windows, power steering.

**Sound system:** Blaupunkt AM/FM-stereo radio/cassette, 4 speakers

### ENGINE
Type .... turbocharged and intercooled 4-in-line, aluminum block and head
Bore x stroke .......... 3.94 x 3.11 in, 100.0 x 78.9mm
Displacement ........................ 151 cu in, 2479cc
Compression ratio ...............................8.0:1
Engine-control system ................... Bosch Motronic
Emissions controls .....3-way catalytic converter, feedback fuel-air-ratio control
Turbocharger ................................ KKK K26
Waste gate ...................................... Porsche
Maximum boost pressure ...................... 10.9 psi
Valve gear .............. belt-driven single overhead cam, hydraulic lifters
Power (SAE net) ................. 217 bhp @ 5800 rpm
Torque (SAE net) ................ 244 lb-ft @ 3500 rpm
Redline ...................................... 6400 rpm

### DRIVETRAIN
Transmission................................... 5-speed

Final-drive ratio ..................... 3.38:1, limited slip

| Gear | Ratio | Mph/1000 rpm | Speed in gears |
|---|---|---|---|
| I | 3.50 | 6.1 | 39 mph (6400 rpm) |
| II | 2.06 | 10.4 | 67 mph (6400 rpm) |
| III | 1.40 | 15.3 | 98 mph (6400 rpm) |
| IV | 1.03 | 20.7 | 132 mph (6400 rpm) |
| V | 0.83 | 25.8 | 153 mph (5950 rpm) |

### DIMENSIONS AND CAPACITIES
Wheelbase ................................... 94.5 in
Track, F/R ............................. 58.1/57.1 in
Length ................................... 166.5 in
Width ...................................... 68.3 in
Height...................................... 50.2 in
Frontal area................................ 20.3 sq ft
Drag coefficient ................................ 0.33
Ground clearance................................ 4.9 in
Curb weight................................. 2850 lb
Weight distribution, F/R (mfr.'s est.)............ 50/50%
Fuel capacity................................. 21.1 gal
Oil capacity ...................................5.8 qt
Water capacity ................................. 9.0 qt

### CHASSIS/BODY
Type ........................... unit construction
Body material ...................welded steel stampings

### INTERIOR
SAE volume, front seat ......................... 50 cu ft
     rear seat ......................... 12 cu ft
     trunk space ......................... 8 cu ft
Front seats .......... bucket (6-way power driver's seat)
Recliner type....................... infinitely adjustable

### SUSPENSION
F: ....... ind, MacPherson strut, coil springs, anti-sway bar
R: ........ind, semi-trailing arm, coil springs, anti-sway bar

### STEERING
Type ..................... rack-and-pinion, power-assisted
Turns lock-to-lock ................................ 3.2
Turning circle curb-to-curb ..................... 33.8 ft

### BRAKES
F:........................ 11.7 x 1.1-in vented disc
R:........................ 11.8 x 0.9-in vented disc
Power assist ...............................vacuum

### WHEELS AND TIRES
Wheel size ............................. F: 7.0 x 16 in; R: 8.0 x 16 in
Wheel type............................. cast aluminum
Tires .............. Pirelli Cinturato P7, F: 205/55VR-16; R: 225/50VR-16
Test inflation pressures, F/R................... 29/36 psi

| MFR.'S PERFORMANCE RATINGS | 944 | 944 Turbo |
|---|---|---|
| Zero to 60 mph | 8.3 sec | 6.1 sec |
| Standing ¼-mile | 16.2 sec | 14.4 sec |
| Top-gear passing, 30–50 mph | 12.5 sec | 10.0 sec |
| 50–70 mph | 12.9 sec | 8.4 sec |
| Top speed | 131 mph | 153 mph |
| Fuel economy, EPA city | 21 mpg | 19 mpg |
| EPA highway | 30 mpg | 26 mpg |

---

systems, upgrading a few critical areas with the substitution of higher-strength materials, and redesigning a few components that simply weren't adequate for life with a turbo. This sounds a lot easier than it actually was. With a bottom-line improvement of nearly 50 percent (217 net horsepower with the turbo, versus 150 net horsepower in normally aspirated trim), Porsche has obviously exceeded the scope of a tack-on turbo kit. What we have here is a basic demonstration of how to design a powerful, fuel-efficient, and clean engine by using the most modern methods.

Internal pressures rise in direct proportion to specific output (at constant rpm), so Porsche substituted new forged-aluminum pistons for the normally aspirated 944's cast pistons. At the same time, the compression ratio was dropped from 9.5 to 8.0:1. The cylinder head uses the same basic casting, machining, and combustion chambers, but it benefits from two key alterations: valve, seat, and guide materials are significantly more heat-resistant, and the exhaust ports contain new ceramic liners. The insulation material delivers a threefold gain: the load on the cooling system is reduced, more exhaust energy is available to aid the turbo's response, and the hotter resulting catalyst-inlet temperatures speed its warmup after a cold start. Another significant change in the cylinder-head area is the use of stiffer valve springs, which increase closing force by twenty percent. Also, oil-pump capacity is up by ten percent. Finally, a new head gasket, featuring a stainless-steel combustion-chamber seal and modified gasket material, has been specified.

These basic engine improvements set the stage for the truly new hardware: a KKK turbocharger, an air-to-air intercooler, and their associated plumbing. The turbo is a water-cooled model K26 manufactured by the German company Kühnle, Kopp, & Kausch. Contrary to popular practice, it is mounted on the intake side of the engine and connected to a stainless-steel-tubing exhaust manifold with a double-walled pipe. The idea is to minimize the detrimental loadings on the turbo while providing both the turbine and the catalyst as much exhaust energy as possible. Toward this end, the manifold is insulated with a fibrous mat held in place by a metal shroud, and the double-walled connecting pipes (manifold to turbine and turbine to catalyst) contain 5mm insulating air gaps. This arrangement reduces turbine-inlet temperature by 160 degrees Fahrenheit, but it also eliminates any chance that excessive heat might be conducted to the turbocharger's bearing housing through heavy-walled manifold castings. Nonetheless, Porsche tests have revealed that the thorough insulation raised the exhaust temperature at the catalyst by 180 degrees and shortened its warmup time by 80 seconds.

The turbocharger's thermal loading is a critical concern because excessive heat—from extended full-load running or from the hot soak that occurs when the car is hastily shut down after a hard run—causes bearing coking (deposits of carbonized oil), which tends to restrict lubricant flow and can in time cause the turbocharger's bearing assembly to fail. Most manufacturers have turned to water cooling to solve the problem, but Porsche has taken this approach one additional step: the 944 Turbo's bearing housing is cooled by flow from the main cooling system while the engine is running, and a temperature-controlled electric water pump takes over to prevent hot-soak overheating when the en-

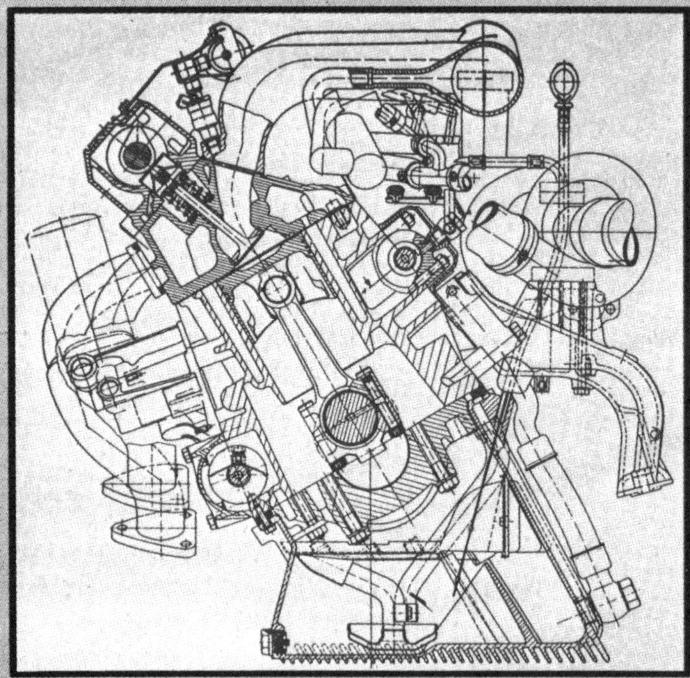

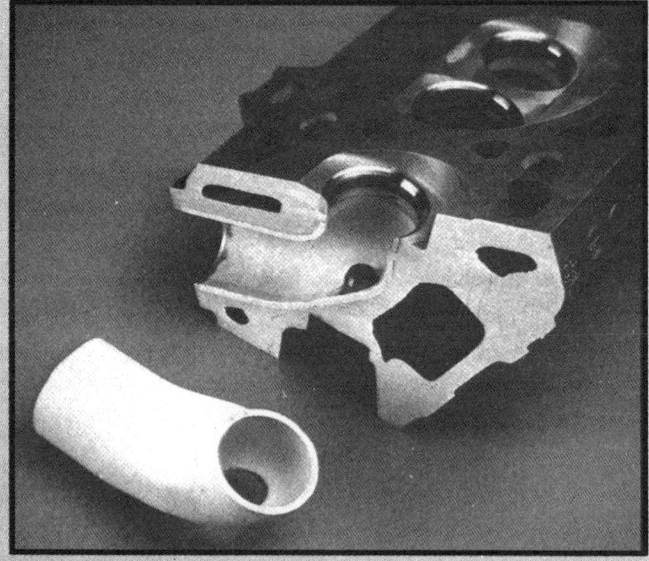

The 944's turbo is located on the intake side for better cooling and for packaging convenience. It's supported by an engine mount, cooled by an electric pump, and fed exhaust by a double-walled pipe. Ceramic exhaust-port liners (above) diminish unwanted heat transfer.

gine is shut off. Thanks to these measures, bearing temperatures never exceed 340 degrees, according to Porsche tests.

Excessive heat is also a concern on the intake side of the engine, so Porsche has given the 944 a generously sized intercooler, a strategy the company introduced to production cars in the 1978 911 Turbo. This 90-square-inch heat exchanger is mounted as far forward as possible and fed air through a 30-square-inch grille opening. The engineers claim a 135-degree drop in charge-air temperature.

In designing the intercooler, the intake manifold, and the tubes that connect them, Porsche had to consider an important trade-off: large tube cross-sections and plenum volumes produce minimal pressure drops and flow losses, but excessively large volumes will also slow the engine's response to a prod of the throttle. A secondary concern is that, below 2200 rpm, there is insufficient exhaust energy to produce useful boost, so the intake-manifold runners have to be long enough to contribute a resonance-effect torque boost. Porsche engineers studied a multitude of combinations before arriving at the final tube and plenum volumes, which provide maximum boost with minimal temperature rise, good low-speed torque, and respectable throttle response.

To help Porsche's turbine and piston engines work happily together, there are three elaborate subsystems—fuel delivery, ignition, and waste gate—all of which are controlled by electronic circuits. A Bosch Motronic computer selects the proper spark advance and air-fuel ratio from an electronic-memory map after studying signals from a variety of sensors. In addition, this computer regulates idle

speed and provides one level of overboost protection: in the event that the intake airflow exceeds a limit value by ten percent for at least three seconds, fuel delivery is interrupted.

Another electronic control circuit is used to manage the turbocharger's waste gate. The system Porsche has adopted here is very similar to Saab's APC (Automatic Performance Control, a design the Swedes unveiled in 1980), in that a solenoid valve is used to modulate the control pressure acting on the waste-gate diaphragm. The solenoid's electronic control circuit monitors both throttle position and a detonation sensor in deciding whether to open or to close the solenoid valve. Like the Motronic computer, the waste-gate controller chooses its boost setting from an electronic map: at full throttle, this system provides maximum low-rpm boost, and then intake-manifold pressure is gradually diminished at higher engine speeds to avoid detonation. The 944 Turbo engine's map has purposely been programmed to deliver a very linear increase in torque as the throttle is depressed.

To maximize cruising fuel economy, the electronic circuit opens the waste gate slightly during part-throttle operation, even when there is no detonation. This reduces exhaust back pressure and forces the throttle to be held further open for a given power output. The net effect is a substantial reduction of pumping losses. To optimize throttle response, the electronic circuit does two things as soon as there is a significant change in the throttle position: the waste gate is quickly closed, directing the full might of the exhaust stream against the turbine wheel, and the steady-speed boost limit is temporarily suspended, help-

ing the engine climb its torque curve.

Since the 944 Turbo's piezoelectric detonation sensor is wired to both the boost-control circuit and the main Motronic computer, engine knock is quelled with combinations of reduced boost and retarded ignition. The system is sophisticated enough to retard the spark only to the cylinder or cylinders that are knocking. Initially, the timing is retarded by three degrees, but it quickly steps back to the normal setting if no more detonation occurs. If knocking persists, the spark can be delayed by a maximum of six degrees.

The bottom line of all this electronic and thermodynamic glory is a whopping 217 net horsepower. This is worldwide output we're talking about: those countries that need a catalyst to produce legal emissions (including the U.S.A.) are given extra boost pressure for global parity. It's another move by the Porsche factory to kill the gray marketeering of its cars.

Porsche engineers are justifiably proud of what they have wrought. The 944 Turbo powerhouse develops an impressive 87.5 horsepower per liter, versus the 911 Carrera's 62.3 hp per liter and the four-valve 928S's 58.1. The power-per-pound figures are equally gratifying: 0.54 hp per pound for the 944 Turbo, versus 0.49 for the 928S.

Now that this particular Porsche is sailing smoothly toward the marketplace with ample power under its belt, it's not hard to imagine what's next. Paul Hensler, Porsche's director of powertrain development, admits that the 928S's four-valve cylinder head will help the 944 Turbo take the next logical step up the technological ladder. . . in less than two years' time.

—*Don Sherman*

# Porsche 928S

## *Faster is better.*

• "The old man would spin in his grave if he saw what they're passing off as a Porsche these days," raved the gas-station attendant. "They've lost the original concept. Where's the rear engine? Where's the air cooling?" Our rebuttal, emphasizing the performance virtues of the 1985-model 928S, failed to sway the man's thinking one iota. To the pump jockey and to many enthusiasts, the only real Porsches are those that perpetuate the design themes of the original 356 model, a car born nearly 40 years ago.

That's a shame, because the good doctor would doubtless be proud of this new 928S. Ferdinand Porsche believed that a sports car must offer transportation superior to a regular sedan's, and this doctrine is a cornerstone of the firm that bears his name today. Sports cars necessarily have smaller payloads than sedans, but for this reason they can be more fuel-efficient and be designed for higher speeds. And, unlike our federal government, Porsche has always understood that, in transportation, faster is better. Jet aircraft have replaced ocean liners and trains for long-distance travel; people drive cars instead of riding bicycles; and even bicycles are faster than they used to be. Faster transportation, simply put, is better transportation.

By the speediness criterion, the 1985 Porsche 928S is a substantial improvement over its predecessor. Top speed is up 10 mph, to 154. Our five-speed test car rocketed from a standing start to 60 mph in 5.7 seconds and hit 100 mph less than eight seconds later; the old car required 6.2 and 17.8 seconds, respectively. And the new model burned through the standing quarter-mile in 14.0 seconds at 102 mph, compared with 14.7 seconds at 94 mph for last year's 928. We also tested the automatic version, and its performance is similarly improved. The automatic's 0-to-60 and quarter-mile times now match those of last year's five-speed, and its top speed is up 10 mph, to 152.

These are amazing figures for a car with extremely tall, fuel-economy-oriented gearing and more than enough power to break the tires loose at low speeds. Credit for the improvements belongs to the new 5.0-liter V-8 engine. With 288 bhp, it's a hefty 54 bhp stouter than the previous 928's 4.7-liter engine, which wasn't what you'd call a weakling. To put this brawn into proper perspective, the new motor makes one-third more power than Chevrolet's fuel-injected V-8 of the same displacement and 104 bhp more than Mercedes' American-specification 5.0-liter V-8.

Even more remarkable than this lofty power output is the Porsche V-8's broadband torque. Although the 302-pound-foot peak occurs at a relatively low 2700 rpm, more than 250 pounds-feet is on hand from 1300 to over 6000 rpm. Most 5.0-liter V-8s do well to exceed that level at any engine speed.

This engine magic is the direct result of the new four-valve-per-cylinder, twin-cam heads (*C/D*, February). The four-valve layout provides good high-rpm breathing, ensuring plenty of power. This in turn allowed Porsche's powertrain engineers to tune the engine's manifolding and valve timing for efficient low- and medium-rpm breathing, providing good torque. The four-valve design also has a pent-roof combustion chamber with a centrally located spark plug, a design that resists detonation and promotes thermodynamically efficient combustion. Consequently, a high compression ratio (10.0:1) could be employed, enhancing both fuel economy and power output at all rpm.

EPA fuel-economy figures have improved slightly from last year's, so the 928S will stay off the dreaded gas-guzzler rolls.

Still, with its 18-mpg over-the-road mileage, this car won't be of much interest to fuel misers. The 928S's efficiency improvements are significant, though, because they are true to the Porsche philosophy that cars are transportation tools. As the company credo goes, fuel efficiency is important to the transportation function and should never be sacrificed, not even for higher performance.

It's good that the firm feels strongly about this, because once a 928S is in the hands of a customer, a touch of the accelerator will wipe out any interest in fuel economy. Not only can the driver effortlessly dial up just about any speed he desires, but the tall gearing and the broad output offer several alternatives to how he goes about his business. To effect nearly instantaneous speed changes, one uses the lower gears, even at very high speeds. Or one can remain exclusively in fifth from about 30 mph and still outpace most traffic. (Since the 928's shift linkage is still not a strong point, this approach is quite attractive.) Or one can specify the four-speed automatic, which always seems ready to transform the engine's plentiful power into blinding speed, requiring only an effortless touch on the accelerator.

Despite the additional camshafts, valves, and output, the new 928S engine is no more obtrusive than its predecessor. This doesn't mean that it's totally isolated from the occupants, for, like most German cars, the 928S doesn't try to deny the fact that it is machinery. The whine of the cam drives, the tap of the hydraulic tappets, the rush of the gases flowing through the intake and exhaust systems, the hum of the various shafts and gears—all blend together to sing the muted song of a happy mechanism.

One would expect this song to be louder when one sees how stuffed with hardware the 928S is. Every nook and cranny under the hood is filled with engine, leading us to speculate that the assembly must somehow be cast in place. The bulky powerplant

forces the front suspension to be positioned so low that it occasionally drags the ground. In back, the spare-tire well, the battery, the transaxle, and a huge muffler occupy every inch of available space. The 928S's spinal column contains a torque tube and big-bore exhaust plumbing.

With so much of its volume full of machinery, it's no wonder that the 928S has little room for its occupants. The front-seat passengers have sufficient space, but anyone banished to the rear seats will want to have some means of forcing the front occupants into close proximity with the dashboard. Luggage space is also in critically short supply.

The two primary occupants are well coddled, however. The all-enveloping, cocoonlike interior is still as attractive in 1985 as it was at the car's introduction in 1978. This year, the major addition is a new set of power seats as standard equipment. Each front bucket has two four-way

switches on its outside flank to control front and rear height, fore-and-aft position, and backrest angle. The bottom cushions are a half-inch lower than before, providing more headroom, and the side bolsters are significantly larger. There is additional lateral support, and it is indeed appreciated, partly because it comes with no loss in comfort. These seats are cushy enough to provide à soft initial sit-down, yet supportive enough for pleasant all-day runs. In short, the 928S's seating is exactly what one expects in a serious transportation tool.

One also expects superior handling and roadholding, and the 928 does not disappoint. No suspension changes were made to accommodate the increased power, because none were necessary: European 928s have been operating at similar power levels for several years, and chassis tuning has been the same here and there for some time. Thus the new 928S has the same uncanny combination of straight-line stability and instant responsiveness that we've come to relish over the years. In everyday driving, its suspension does a nice job of communicating pertinent tire-to-road information while blocking the transmittal of most pavement imperfections. When the road gets seriously twisty, the 928S seems to become smaller and lighter on its feet. It's supremely controllable with either the

steering or the throttle in hard corners, and any decent driver can use its impressive 0.83-g grip with confidence and safety.

The braking system is also reassuring, thanks to linear performance, well-proportioned front-to-rear balance, and excellent modulation. These characteristics contribute to the 928S's ability to stop from 70 mph in just 175 feet. Perhaps even more impressive is the brake system's ability to absorb the energy of triple-digit speeds without fading or emitting any disconcerting squeals, groans, or odors. The only thing missing is the anti-lock system standard on European 928Ss.

In the design of a versatile, high-speed transportation device, no aspect of performance can be overlooked. Without its strong brakes, for example, the 928's strong engine could not be fully exploited to provide faster and better transportation. Dr. Porsche espoused this attitude, and we think he would be well satisfied with the way the 928S follows his philosophies— even if it does have a water-cooled, front-mounted engine.

Unfortunately, 40 years of progress does cost a lot: the new 928S will set you back a cool $50,000. But if your taste in very fast cars runs toward the functional rather than the spectacular, you'll be hard pressed to find a better deal than the one available at your neighborhood Porsche store.

—*Csaba Csere*

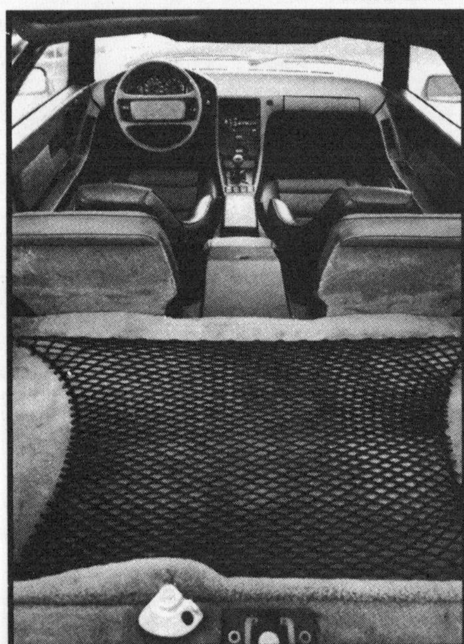

• I can tell you what this car is like in the snow. It's *hilarious!*

I had no chance to run our 928 in dry weather. Now the weather is all white, and I'm going out of town, and the meanies at Porsche are hand wringing about getting their four-valve 928S toy back pronto. So it was me and the four-valve against snow white. The four-valve and I won in a sideways breeze.

See, we've got the automatic here, and that takes all the slack out of it. The engine behaves so beautifully at idle that you simply release the brakes and you get rolling pretty as you please. The revs rise so controllably that you make delicate adjustments without distraction, so your attention goes farther up the road. So well distributed is the weight of the 928 that it even accelerates with a rush in the snow. And when it yaws, the softest of dabs with the wheel sets you straight—amazing for a car with wide, flat VR-rated tires and enough power to put Con Ed on the skids. Only the mighty brakes need extra-special care.

I want my toy back when the weather is good. The boys in blue may not think so, but the hilarity has just begun.

—*Larry Griffin*

Since I'm the only bloke around here who's been lucky enough to drive both a five-speed and an automatic 928S, the difficult job of picking between the two falls on my shoulders. This used to be easy for me because the manual gearbox had so many flaws. The automatic was made for the job because it minimized the interruptions of power flow from the lovely V-8 engine.

Porsche apparently knew that its five-speed wasn't perfect: even though only a thousand or so manual 928s are sold in the U.S. each year, the engineers have redesigned certain parts of the shift linkage in order both to shorten the throws and to reduce the effort. (Fans of the "racing" five-speed shift pattern will be happy to hear that first gear is still down and to the left, off the H.)

So far I've seen two 1985 928S five-speeds that shifted beautifully (those at the initial preview) and one that was cantankerous (our test car). In the last case, second gear was vague and hard to find, the clutch engagement felt nonlinear, and the shifter's spring loading to the right of the H was annoyingly high. Until I have the chance to conduct a more comprehensive survey, my recommendation stays with the automatic.

—*Don Sherman*

I'm still thrilled at the sight of a 928, S or otherwise, coming up in my rear-view mirror. There is no current-production Ferrari or Lamborghini that could give me greater pleasure than a 32-valve 928S, and there is no car I would rather own. It is unfortunate that the 928S of my dreams would cost me about what I would pay for a farm in the Irish Hills, which I can't afford either. I'd be torn on the subject of automatic versus five-speed manual gearbox. In my heart I know that all Porsches should be equipped with five-speed manuals, and I love the five-speed that comes with the 928S. But, on the other hand, I also know that the Porsche 928S engine, with its eight cylinders, 32 valves, 288 horsepower, and 302 pounds-feet of torque, needs a manual transmission about as much as Susan Sarandon needs a third nostril. Maybe I'd get the automatic and try to drive it so that onlookers would think it had the manual. I dunno. I *do* know, however, that this car embodies everything, inside and out, that I regard as the embodiment of automotive virtue. There is, for me, simply none nicer. —*David E. Davis, Jr.*

**Vehicle type:** front-engine, rear-wheel-drive, 2+2-passenger, 3-door coupe

**Price as tested:** $50,000

**Options on test car:** none

**Sound system:** Blaupunkt Köln AM/FM-stereo radio/cassette, 8 speakers, 9 watts per channel

## ENGINE
Type .................... V-8, aluminum block and heads
Bore x stroke ......... 3.94 x 3.11 in, 100.0 x 78.9mm
Displacement ..................... 303 cu in, 4957cc
Compression ratio ..................... 10.0:1
Fuel system ............ Bosch LH-Jetronic fuel injection
Emissions controls.....3-way catalytic converter, feedback fuel-air-ratio control, auxiliary air pump
Valve gear .. belt- and chain-driven double overhead cams, 4 valves per cylinder, hydraulic lifters
Power (SAE net)................ 288 bhp @ 5750 rpm
Torque (SAE net)............. 302 lb-ft @ 2700 rpm
Redline ............................... 6100 rpm

## DRIVETRAIN (5-speed)
Final-drive ratio ..................... 2.20:1

| Gear | Ratio | Mph/1000 rpm | Max. test speed |
|---|---|---|---|
| I | 4.07 | 8.1 | 49 mph (6100 rpm) |
| II | 2.72 | 12.1 | 74 mph (6100 rpm) |
| III | 1.93 | 17.0 | 104 mph (6100 rpm) |
| IV | 1.46 | 22.5 | 137 mph (6100 rpm) |
| V | 1.00 | 32.8 | 154 mph (4700 rpm) |

## DRIVETRAIN (automatic)
Final-drive ratio ..................... 2.20:1

| Gear | Ratio | Mph/1000 rpm | Max. test speed |
|---|---|---|---|
| I | 3.68 | 8.9 | 54 mph (6100 rpm) |
| II | 2.41 | 13.6 | 83 mph (6100 rpm) |
| III | 1.44 | 22.8 | 139 mph (6100 rpm) |
| IV | 1.00 | 32.8 | 152 mph (4650 rpm) |

## DIMENSIONS AND CAPACITIES
Wheelbase .................................. 98.4 in
Track, F/R ........................... 60.4/59.9 in
Length .................................. 175.7 in
Width ...................................... 72.3 in
Height ..................................... 50.5 in
Frontal area ............................. 21.0 sq ft
Ground clearance.......................... 4.7 in
Curb weight (5-speed) .................... 3450 lb
Weight distribution, F/R (5-speed) .........51.0/49.0%
Fuel capacity ............................ 22.7 gal
Oil capacity ............................... 7.9 qt
Water capacity ...........................16.0 qt

## CHASSIS/BODY
Type ........................... unit construction
Body material ...... welded steel and aluminum stampings

## INTERIOR
SAE volume, front seat .................... 53 cu ft
rear seat ..................... 21 cu ft
trunk space .................... 8 cu ft
Front seats ............. bucket with 8-way power assist
Recliner type .................... infinitely adjustable
General comfort ...................poor fair good **excellent**
Fore-and-aft support ..............poor fair good **excellent**
Lateral support .................poor fair good **excellent**

## SUSPENSION
F:.................... ind, unequal-length control arms, coil springs, anti-sway bar
R:.................... ind, unequal-length control arms, coil springs, anti-sway bar

## STEERING
Type .................... rack-and-pinion, power-assisted
Turns lock-to-lock ......................... 3.0
Turning circle curb-to-curb ................... 37.7 ft

## BRAKES
F:............................. 11.1 x 1.3-in vented disc
R:............................. 11.4 x 0.8-in vented disc
Power assist ...............................vacuum

## WHEELS AND TIRES
Wheel size ......................... 7.0 x 16 in
Wheel type....................... forged aluminum
Tires .......... Dunlop SP Sport Super D4, 225/50VR-16
Test inflation pressures, F/R ................. 36/42 psi

# CAR AND DRIVER TEST RESULTS

| ACCELERATION (5-speed) | Seconds |
|---|---|
| Zero to 30 mph .......................... | 2.3 |
| 40 mph ................... | 3.1 |
| 50 mph ................... | 4.4 |
| 60 mph ................... | 5.7 |
| 70 mph ................... | 7.2 |
| 80 mph ................... | 9.2 |
| 90 mph ................... | 11.3 |
| 100 mph ................... | 13.5 |
| Top-gear passing time, 30–50 mph .......... | 8.9 |
| 50–70 mph .......... | 9.7 |
| Standing ¼-mile............. 14.0 sec @ 102 mph | |
| Top speed ................... | 154 mph |

| ACCELERATION (automatic) | Seconds |
|---|---|
| Zero to 30 mph .......................... | 2.5 |
| 40 mph ................... | 3.5 |
| 50 mph ................... | 4.6 |
| 60 mph ................... | 6.2 |
| 70 mph ................... | 7.9 |
| 80 mph ................... | 9.8 |
| 90 mph ................... | 12.9 |
| 100 mph ................... | 17.1 |
| Top-gear passing time, 30–50 mph .......... | 3.5 |
| 50–70 mph .......... | 3.5 |
| Standing ¼-mile............. 14.7 sec @ 94 mph | |
| Top speed ................... | 152 mph |

**BRAKING**
70–0 mph @ impending lockup.............. 175 ft
Modulation....................poor fair good **excellent**
Fade ......................... **none** moderate heavy
Front-rear balance .................... poor fair **good**

**HANDLING**
Roadholding, 300-ft-dia skidpad .......... 0.83 g
Understeer................... **minimal** moderate excessive

**COAST-DOWN MEASUREMENTS**
Road horsepower @ 50 mph ............... 15.5 hp
Friction and tire losses @ 50 mph ........... 6.5 hp
Aerodynamic drag @ 50 mph ............... 9.0 hp

**FUEL ECONOMY**
EPA city driving, 5-speed ............... 16 mpg
automatic ............... 17 mpg
EPA highway driving, 5-speed ........... 25 mpg
automatic ............... 23 mpg
C/D observed, 5-speed ................... 18 mpg

**INTERIOR SOUND LEVEL (5-speed)**
Idle ................................... 55 dBA
Full-throttle acceleration.................... 76 dBA
70-mph cruising ........................ 74 dBA
70-mph coasting ........................ 74 dBA

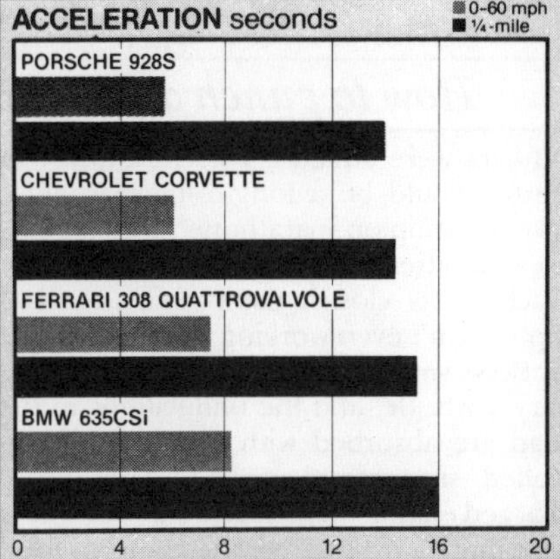



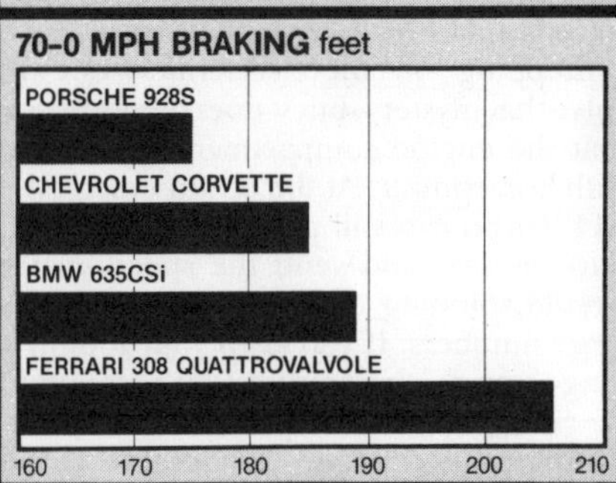





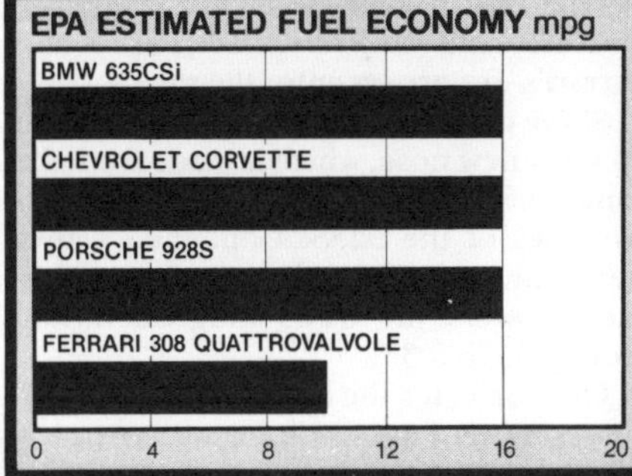



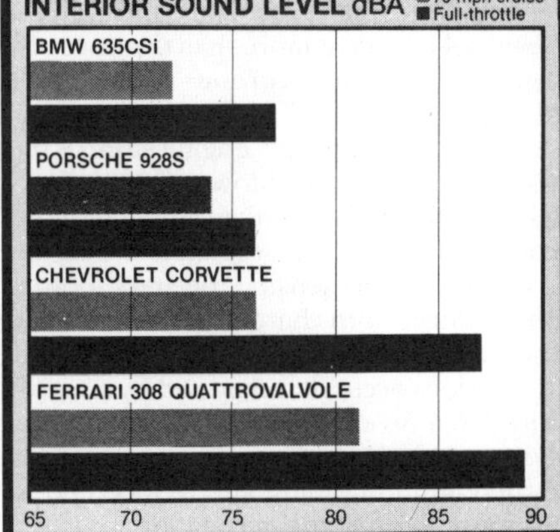

ROAD TEST

# Porsche 944 Turbo

*How to punch a 157-mph hole in the atmosphere.*

• If cars were athletes, Porsche's new 944 Turbo would be a long-distance runner. Like a champion marathoner, this car eats up mile after mile at speeds beyond the reach of its closest competitors. At 100 mph, it isn't even working up a sweat. The air flows smoothly over the car, generating nary a whistle, and the undulations in the road are absorbed with supple and controlled suspension motion. The turbocharged engine hums such a smooth, sweet tune that it's hard to tell whether it has four, six, eight, or even twelve cylinders.

Stepping on the accelerator doesn't solve that mystery, but it does demonstrate that the engine compartment is bursting with horsepower. At the century mark, the 944 Turbo can still push the driver deep into his seat and send the speedometer needle spinning dizzily toward magically large numbers. If you keep your foot in it long enough, the pointer won't stop until it reaches 157 mph. Such speed distinguishes the four-cylinder 944 Turbo as the fastest car sold in America— at least until the new 911 Turbo and Ferrari's Testarossa enter the race.

Some of the Turbo's speed is a result of its sleek new nose, which is made from the same deformable plastic used on the extremities of the 928S. Combined with a new rear under-tray, the new front treatment lowers the 944's drag coefficient from 0.35 to 0.33.

Of course, it's the turbo engine that deserves most of the credit for adding nearly 30 mph to the top speed of the 944. The blown powerplant develops 217 bhp, a healthy 47 percent more than the normally aspirated engine. Torque is improved even more, from 137 to 244 pounds-feet.

Obviously, Porsche's engine wizards did more than just add a blower to their four-banger. They also fitted an air-to-air intercooler ahead of the radiator, which increases charge density, bolsters power, and reduces both charge temperature and the likelihood of engine knock. When knock does occur, it is detected by a sensor wired to a Bosch Motronic engine-control system, which uses combinations of retarded ignition timing and reduced boost pressure to quell the detonation.

To help the engine live through the stresses imposed by its 10.9-psi peak boost pressure, the Porsche engineers reduced the 944's compression ratio from 9.5:1 to 8.0:1 and specified forged pistons, stiffer valve springs, high-temperature-resistant valves and valve seats, a stainless-steel head gasket, and a larger-capacity oil pump. Such changes are normal design practice for turbo motors. The 944 Turbo also sports ceramic exhaust-port liners and double-walled and insulated stainless-steel tubing to pipe the exhaust gases to the remote-mounted KKK turbocharger. These features reduce the heat transferred from the turbocharger back into the engine while improving catalyst light-off time. The water-cooled turbocharger even has its own cooling circuit and pump, which keep its bearing temperatures within prescribed limits. Such measures ensure that catalyst-equipped 944 Turbos will survive sustained autobahn running (and also hasty shutdowns thereafter) in the German market.

Not surprisingly, the four-cylinder's newfound power gives the 944 Turbo acceleration to match its impressive top speed. Despite its 3040-pound curb weight (about 200 pounds more than a normally aspirated 944's), the Turbo rockets from a standstill to 60 mph in six seconds flat and shoots through the standing quarter-mile in 14.5 seconds at 97 mph; the last normally aspirated 944 we tested hit 60 mph in 7.4 seconds and the quarter-mile mark in 15.5 seconds at 87 mph. As their speeds increase, the gap between the two cars grows rapidly. From 0 to 100 mph, the Turbo needs only 15.6 seconds, versus 21.7 for the standard car; to 120, the figures are 23.6 and 49.5, respectively. If you're after the longest set of legs you can get, take your 944 *with* the turbo.

That's not to say that everything is better with the new powertrain. In low-speed driving, for example, the Turbo feels more muscle-bound than powerful. Unless you punish either the tires or the clutch by starting hard enough to keep the turbo on full boil, the boosted 944 feels sluggish off the line; flooring the throttle after a normal clutch engagement produces little re-

sponse for at least a second. And in top gear, the Turbo requires 14.7 seconds to accelerate from 30 to 50 mph, versus 12.0 for the standard car. (The Turbo is quicker in our 50-to-70-mph fifth-gear test.)

The Turbo's lackluster low-speed performance is quite understandable in view of its lower compression ratio, greater weight, and taller gearing. (Fifth has about the same ratio on both the turbocharged and the normally aspirated car, but the Turbo's intermediate gears are from fifteen to eighteen percent taller.) The result is considerably less snap until the boost comes in. The Turbo definitely lacks the instant reflexes of a Corvette or a 911.

In its favor, however, the 944 Turbo is one of the most livable 150-mph-plus cars we've driven. We averaged an impressive 22 mpg during our testing and found the

Turbo to be a very comfortable traveling companion. Wind ruffle and mechanical noises are subdued, and the Turbo rides better than older 944s. All 1986-model 944s have new forged-aluminum control arms locating their front struts instead of the modified VW Rabbit parts that were used on previous 944s. The new pieces have larger bushings that are more compliant over small, sharp bumps. Our Turbo seemed to pound less over such imperfections than any other 944 we've driven.

Another change for all 944s is the new dashboard. The design is more flowing and open than the old dash and has a four-

dial instrument cluster similar to that of the 928S. The dials have white-on-black markings instead of the annoying yellow graphics of the old car. Some of the knobs and switches look a bit insubstantial for a $30,000 car, but the basic layout is logical and clean. The Porsche designers went so far as to hide the trip odometer's reset button in one of the airflow registers.

Finding it may not be easy, because there are enough airflow registers to cover about half of the new dash. In conjunction with a higher-capacity air-conditioning system, they can produce an arctic chill quickly, without howling gale-force winds. The new dash also has a more convenient radio location and a larger glove box.

To upgrade the furniture for this new version of the 944, Porsche simply borrowed the seats from the current 928S. They are similar in comfort and support to the old 944 non-sport seats, and they have a very accessible ratchet-type backrest adjuster located under the front edge of the lower cushion. (The old seats had knobs on the side that were hard to reach when the doors were shut.) The Turbo's seats

# To the Track and Back

*Wherein racing improves the breed—at least a little bit.*

• Kurt Wöhr is standing in the back of a long white trailer crammed full of aluminum air-freight containers that have come straight from Germany, and he is handing out parts. Porsche 944 Turbo parts. Wöhr is here in the Lime Rock Park paddock to supply the Porsche Showroom Stock racers, who are trying to break the Corvette's stranglehold on the SCCA/*Playboy* United States Endurance Cup series. The only incongruity is that Wöhr isn't from Porsche's racing department. Rather, he's a project supervisor from the production-line side of Porsche engineering. Welcome to research and development on the run.

No other car company has more of its ego or image wrapped up in racing than Porsche. The wizards of Weissach have their fingers into every motorsports pie

from Formula 1 to Showroom Stock. Since Porsche specializes in sports cars, the racing tie-in would seem natural. Unfortunately, it's not so black and white.

Porsche, like any company deeply involved in motorsports, wants you to believe that racing improves the breed. Truth be known, little if anything developed on the track ever gets into production models—Porsche's or anyone else's. But there are exceptions, and the 944 Turbo is one of them. It was actually christened on a racetrack, and it's been developed under fire ever since.

The racing history of the 944 Turbo begins at the 24 Hours of Le Mans in 1981. The normally aspirated 944 hadn't even debuted yet, but here was Porsche with a competition version. Though it was referred to as a 924, this special racer

used what was to become the 944's engine block. It was topped off with a sixteen-valve cylinder head (still not in production for the 944) and a turbocharger, which together pumped its power up to 410 hp. "We were testing some concepts" is all Wöhr will say about it. The car was never raced again.

Since then, all of the 944 Turbo's preproduction racing miles were rolled up in the SCCA Showroom Stock series. "This is to help us test and develop the car," explains Wöhr in his thick German accent. "We want a normal production car to be able to drive like this for 24 hours without any extra work being done by the owner."

But Porsche has its own race course at Weissach, along with all the drivers it could ever need to do flat-out 24-hour

also have power front and rear height adjusters, which combine with a slightly higher and less vertical steering wheel to solve the old car's thigh-clearance problem.

With its improved comfort and prodigious speed, the 944 Turbo should be an excellent long-distance cruiser. Unfortunately, our test car was not, because it didn't track very well. The steering felt dead on center, and the car constantly wandered from straight ahead. On smooth roads, frequent minor corrections were needed to keep the Turbo on a chosen path, and over undulating pavement at high speed, the car was all over the road. Previous 944s we've driven did not exhibit these problems, but none of them was fitted with Pirelli P7 tires. Improper wheel alignment could also account for such behavior, but since our car had been freshly serviced at Porsche North America headquarters in Reno, Nevada, that explanation seems unlikely. Until we try another example, we'll have to withhold our final blessing on the Turbo's steering.

In other handling respects, the Turbo feels much like previous 944s—which is to say, outstanding. In hard cornering, the power steering responds well and provides a direct link to the tires' activities. The chassis feels sure-footed over all manner of bumps and keeps its tires firmly planted on the pavement at all times. Mild understeer is the Turbo's style at the limit, but it can be changed to benign oversteer by progressively lifting off the throttle.

Although the Turbo shares many handling traits with the standard 944, it doesn't feel quite as nimble as the lighter car. It feels locked to the road most of the way up to its 0.80-g cornering limit. Previous 944s tended to slide more easily, a characteristic which made them more "chuckable," and also more endearing to the *C/D* staff.

Of course, with the Turbo's bigger brakes, there's little reason to toss it sideways into corners. The finned aluminum calipers and thick vented discs brought our test car to a stop from 70 mph in only 186 feet, despite a bit of early rear lockup. Moreover, not even hard use from 150 mph causes noticeable fade, pained squeals, or vibration through the pedal.

Clearly, the Porsche engineers have done their homework, creating yet another car that is not only fast but also well rounded. By the numbers, the 944 Turbo appears to be very competitive with the 911 Carrera and the 928S. The Turbo is, of course, the junior member of this elite trio, and so it makes its driver work much harder to generate the straight-line performance that the others produce effortlessly.

On the other hand, the Turbo delivers much of the 928S's comfort and refinement for about $20,000 less. And it demands less skill to drive quickly than the slightly more expensive Carrera. That's a combination of qualities that should snare plenty of buyers for Porsche's newest high-speed wonder. —*Csaba Csere*

runs. "Ah," says Wöhr, shaking his finger, "*that* is not as hard as a race. Over here on a hot day in a race, we get much higher temperatures in the engine and brakes than in Germany. When something breaks, we go back and have a meeting and decide how to fix it for production."

Virtually all of the 944 Turbo's early U.S. racing development was done in conjunction with veteran racer Freddy Baker, a Porsche dealer in Bedford, Ohio. The Turbo made its first appearance in the U.S. in 1983, in the prototype class at the Nelson Ledges 24-hour Showroom Stock race. Porsche used this outing to test the effects of running on unleaded fuel with a catalytic converter for 24 flat-out hours. No problems there, but others did crop up.

"We were running in first place," says Baker. "Then a ball joint broke." The engineers had been trying to sell management on sturdier suspension components, and this mishap swung the balance. The forged control arms you see on 944s today are the result.

"We suffered some loss in turbo power during the race," Baker goes on. Changes were made to ensure that full boost was always available. "The gas tank split, so it was changed from steel to plastic, and it was enlarged. They also redesigned the header pipe."

It was a full year before the 944 Turbo came out of the shadows again. Once more the venue was Nelson Ledges, and the car was again entered as a prototype. "We had a perfect run," says Baker. "We won. So all we did was change the oil and take it to the 24-hour race at Mid-Ohio. After six hours, the engine blew." This misfortune led to the strengthening of the powerplant's bottom end.

At this year's running of the Longest Day of Nelson, Baker's 944 Turbo finished second in the prototype class to a Corvette. That's the best a 944 Turbo has done this year.

"We have less low-end torque than the Corvettes," complains Baker, "and worse fuel economy." But the biggest problem is tires. "The Corvettes have 255s, but we have only 205s on the front and 225s on the rear. What can we do?"

The SCCA has already okayed a switch to the larger, 225-section rubber up front, but it hasn't helped much. You better believe that the engineers back at Weissach are scrambling to stuff even larger tires into the tight wheel wells.

Getting competitive is what it's all about to the racers, but for Porsche, as we said, it isn't so black and white. "Sure, developing the car is important," explains Peter Schmitz, assistant director of Porsche Motorsport of North America, the company's U.S. racing arm. "But let's face it. Racing is good public relations."

For this company, good PR means only one thing: winning. If the 944 Turbo doesn't start doing some of that pretty soon, you can bet your alloy wheels that R&D on the run will come to a screeching halt. —*Rich Ceppos*

**Vehicle type:** front-engine, rear-wheel-drive, 2+2-passenger, 3-door coupe

**Price as tested:** $31,398

**Options on test car:** base Porsche 944 Turbo, $29,500; power sunroof, $695; AM/FM-stereo radio/cassette, $625; freight, $578

**Standard accessories:** power steering, windows, and seats, A/C, rear defroster

**Sound system:** Blaupunkt Monterey AM/FM-stereo radio/cassette, 4 speakers

### ENGINE

Type . . . . turbocharged and intercooled 4-in-line, aluminum block and head
Bore x stroke . . . . . . . . . 3.94 x 3.11 in, 100.0 x 78.9mm
Displacement . . . . . . . . . . . . . . . . . . . 151 cu in, 2479cc
Compression ratio . . . . . . . . . . . . . . . . . . . . . . . 8.0:1
Engine-control system . . . . . . . . . . . . . . Bosch Motronic with port fuel injection
Turbocharger . . . . . . . . . . . . . . . . . . . . . . . . . . . . . . KKK
Waste gate . . . . . . . . . . . . . . . . . . . . . . . . . . . . Porsche
Maximum boost pressure . . . . . . . . . . . . . . . . . 10.9 psi
Valve gear . . . . . . . . . . . . . belt-driven single overhead cam, hydraulic lifters
Power (SAE net) . . . . . . . . . . . . . . . 217 bhp @ 5800 rpm
Torque (SAE net) . . . . . . . . . . . . . . 244 lb-ft @ 3500 rpm

### DRIVETRAIN

Transmission . . . . . . . . . . . . . . . . . . . . . . . . . . . 5-speed
Final-drive ratio . . . . . . . . . . . . . . . . 3.38:1, limited slip

| Gear | Ratio | Mph/1000 rpm | Max. test speed |
|---|---|---|---|
| I | 3.50 | 6.1 | 39 mph (6400 rpm) |
| II | 2.06 | 10.4 | 67 mph (6400 rpm) |
| III | 1.40 | 15.3 | 98 mph (6400 rpm) |
| IV | 1.03 | 20.8 | 133 mph (6400 rpm) |
| V | 0.83 | 25.8 | 157 mph (6100 rpm) |

### DIMENSIONS AND CAPACITIES

Wheelbase . . . . . . . . . . . . . . . . . . . . . . . . . . . . 94.5 in
Track, F/R . . . . . . . . . . . . . . . . . . . . . . . . 58.1/57.1 in
Length . . . . . . . . . . . . . . . . . . . . . . . . . . . . . 166.5 in
Width . . . . . . . . . . . . . . . . . . . . . . . . . . . . . . . 68.3 in
Height . . . . . . . . . . . . . . . . . . . . . . . . . . . . . . 50.2 in
Ground clearance . . . . . . . . . . . . . . . . . . . . . . . 4.9 in
Curb weight . . . . . . . . . . . . . . . . . . . . . . . . . 3040 lb
Weight distribution, F/R . . . . . . . . . . . . . . 50.7/49.3%
Fuel capacity . . . . . . . . . . . . . . . . . . . . . . . . 21.1 gal

### CHASSIS/BODY

Type . . . . . . . . . . . . . . . . . . . . . . . . unit construction
Body material . . . . . . . . . . . . . . welded steel stampings

### INTERIOR

SAE volume, front seat . . . . . . . . . . . . . . . . 50 cu ft
rear seat . . . . . . . . . . . . . . . . . . . . 12 cu ft
trunk space . . . . . . . . . . . . . . . . . 12 cu ft
Front seats . . . . . . . . . . . . . . . . . . . . . . . . . . bucket
Seat adjustments . . . . . . . fore and aft, seatback angle, front height, rear height
General comfort . . . . . . . . . . . . . . poor fair good **excellent**
Fore-and-aft support . . . . . . . . . . . poor fair good **excellent**
Lateral support . . . . . . . . . . . . . . . poor fair **good** excellent

### SUSPENSION

F: . . . . . . . . ind, strut located by a control arm, coil springs, anti-roll bar
R: . . . . . . . . ind, semi-trailing arm, coil springs, anti-roll bar

### STEERING

Type . . . . . . . . . . . . . . . . . . rack-and-pinion, power-assisted
Turns lock-to-lock . . . . . . . . . . . . . . . . . . . . . . . . 3.2
Turning circle curb-to-curb . . . . . . . . . . . . . . . 33.8 ft

### BRAKES

F: . . . . . . . . . . . . . . . . . . . . . 11.7 x 1.1-in vented disc
R: . . . . . . . . . . . . . . . . . . . . . 11.8 x 0.9-in vented disc

### WHEELS AND TIRES

Wheel size . . . . . . . . . . . . . . . . F: 7.0 x 16 in; R: 8.0 x 16 in
Tires . . . . . . . . . . . . . . Pirelli Cinturato P7, F: 205/55VR-16; R: 225/50VR-16

## CAR AND DRIVER TEST RESULTS

### ACCELERATION

| | Seconds |
|---|---|
| Zero to 30 mph | 2.1 |
| 40 mph | 3.1 |
| 50 mph | 4.6 |
| 60 mph | 6.0 |
| 70 mph | 8.1 |
| 80 mph | 10.3 |
| 90 mph | 12.5 |
| 100 mph | 15.6 |
| 110 mph | 19.4 |
| Top-gear passing time, 30–50 mph | 14.7 |
| 50–70 mph | 10.6 |
| Standing ¼-mile | 14.5 sec. @ 97 mph |
| Top speed | 157 mph |

### BRAKING

70–0 mph @ impending lockup . . . . . . . . . . . . . 186 ft
Modulation . . . . . . . . . . . . . . . poor fair **good** excellent
Fade . . . . . . . . . . . . . . . . . **none** moderate heavy
Front-rear balance . . . . . . . . . . . . . . poor **fair** good

### HANDLING

Roadholding, 300-ft-dia skidpad . . . . . . . . . . . . 0.80 g
Understeer . . . . . . . . . . . . . **minimal** moderate excessive

### COAST-DOWN MEASUREMENTS

Road horsepower @ 30 mph . . . . . . . . . . . . . . . 7 hp
50 mph . . . . . . . . . . . . . . . 14 hp
70 mph . . . . . . . . . . . . . . . 27 hp

### FUEL ECONOMY (projected)

EPA city driving . . . . . . . . . . . . . . . . . . . . . 19 mpg
EPA highway driving . . . . . . . . . . . . . . . . . . 24 mpg
*C/D* observed fuel economy . . . . . . . . . . . . . 22 mpg

### INTERIOR SOUND LEVEL

Idle . . . . . . . . . . . . . . . . . . . . . . . . . . . . . . . 57 dBA
Full-throttle acceleration . . . . . . . . . . . . . . . 77 dBA
70-mph cruising . . . . . . . . . . . . . . . . . . . . . 69 dBA
70-mph coasting . . . . . . . . . . . . . . . . . . . . . 69 dBA

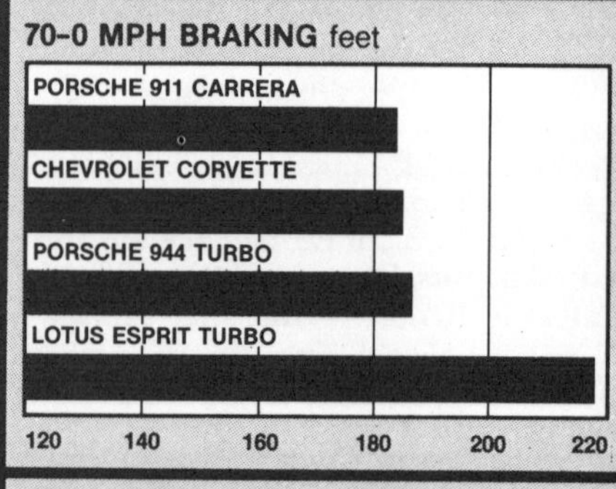

**CURRENT BASE PRICE** dollars x 1000

- LOTUS ESPRIT TURBO
- PORSCHE 911 CARRERA
- PORSCHE 944 TURBO
- CHEVROLET CORVETTE

(scale: 0, 12, 24, 36, 48, 60)

**ACCELERATION** seconds
(legend: 0–60 mph; ¼-mile)

- PORSCHE 911 CARRERA
- CHEVROLET CORVETTE
- PORSCHE 944 TURBO
- LOTUS ESPRIT TURBO

(scale: 0, 3, 6, 9, 12, 15)

**70–0 MPH BRAKING** feet

- PORSCHE 911 CARRERA
- CHEVROLET CORVETTE
- PORSCHE 944 TURBO
- LOTUS ESPRIT TURBO

(scale: 120, 140, 160, 180, 200, 220)

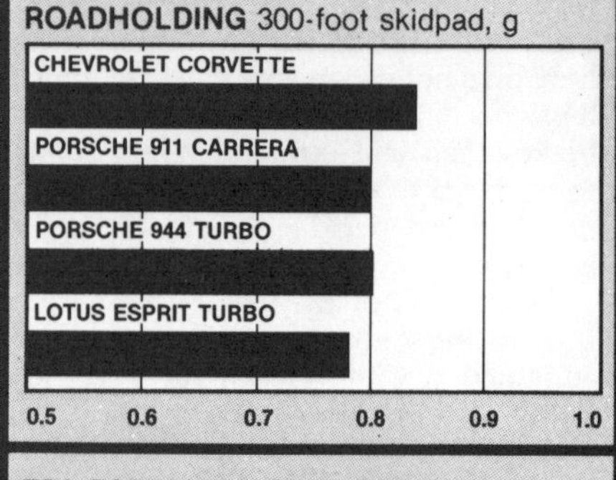

**ROADHOLDING** 300-foot skidpad, g

- CHEVROLET CORVETTE
- PORSCHE 911 CARRERA
- PORSCHE 944 TURBO
- LOTUS ESPRIT TURBO

(scale: 0.5, 0.6, 0.7, 0.8, 0.9, 1.0)

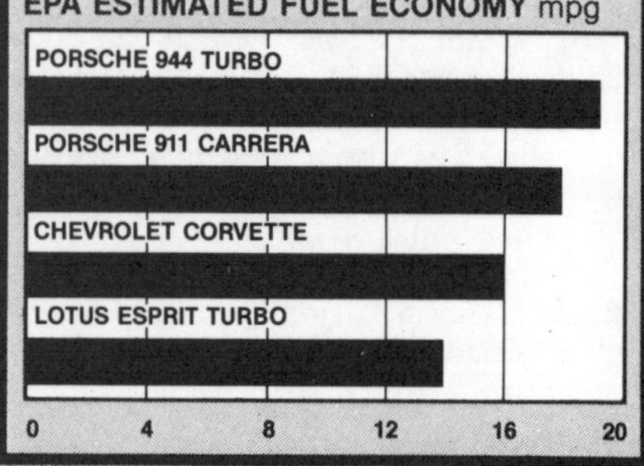

**EPA ESTIMATED FUEL ECONOMY** mpg

- PORSCHE 944 TURBO
- PORSCHE 911 CARRERA
- CHEVROLET CORVETTE
- LOTUS ESPRIT TURBO

(scale: 0, 4, 8, 12, 16, 20)

# Ruf Porsche Turbo

*The designer Porsche for those whose speed ambitions go beyond appearances.*

• Designer cars are all the rage these days. An entire cottage industry has sprung up to meet the individualistic automotive needs of the world's well-heeled customers. Much of this business focuses on cosmetic alterations that attempt to project the buyers' aggressive self-images. Other companies concentrate on functional changes to tailor a car to a customer's particular driving needs.

An example of the latter is Ruf, located near Munich, West Germany, specializing in modified Porsche 911s and Turbos. Unlike many other Porsche *couturiers,* the Ruf people eschew sloped noses, outrageous fender flares, and triple-decker spoilers. Performance is their business, and they go after it with few compromises.

Ruf is serious, indeed, judging by the European-specification, Ruf-modified Turbo we drove recently. This car shimmers with motivational energy, packing enough power to transform a long straight into a short chute, to turn a corner into a launch ramp for the next rocketlike linear thrust. A semicontrolled inferno under the deck lid spits and cracks as it escapes through the exhaust. This hot Porsche accelerates to 60 mph in 4.8 seconds and through the quarter-mile in 13.2 seconds at 107 mph, and it doesn't stop clawing for speed until it reaches 169 mph.

The source of this highly charged mobility is a Ruf-modified Porsche engine pumped up from 290 to 360 bhp. Instead of crudely increasing boost and compromising reliability, Ruf engineers reworked the factory motor extensively for more high-rpm power. The modifications include ported cylinder heads, more radical camshafts, a higher-flow turbocharger, a larger intercooler, a less restrictive muffler, and bigger-bore pistons, which enlarge the displacement from 3299 to 3367cc.

To make the best of the loftier but narrower power curve, the Ruf people also fitted a five-speed transaxle of their own design, which evolved from the standard four-speed Porsche unit. In addition to the new gear, the five-speed provides closer gear spacing, a shorter first, and a taller top ratio than the standard gearbox offers.

Taller gearing notwithstanding, the Ruf car is happiest at high rpm. Although low-speed operation is smooth and tractable, full throttle produces only modest acceleration until about 3500 rpm, when the car bursts forward as if shot from a cannon. For instant response, the Ruf driver must work his shift lever diligently to keep the turbo near full boil. Fortunately, this is no great hardship, for the Ruf gearbox ratios are well placed, and the shifter is precise and satisfying.

The Ruf Turbo's handling makes fewer demands on the driver than a Porsche Turbo does, which is surprising in view of the power on tap. Ruf has tamed the Porsche's traditional trailing-throttle oversteer so well that the car remains remarkably unruffled until the last increment of its 0.87-g lateral adhesion is exploited. The usual tail-out tendencies do crop up then, whether you use too little or too much throttle. All in all, though, this is one of the most docile rear-engined Porsches we've ever encountered.

Much of the credit for the Ruf's friendly handling must go to the Yokohama A-008 tires fitted to our test car. The front tires are the usual 205/55VR-16s, but the rears are up from the usual 225/50VR-16s to 245/45VR-16s. (Rumor has it that this size will be used at the rear of the 1986 Porsche Turbo and 928S.) The Ruf Porsche clearly likes the additional rear grip. It also benefits from Ruf-calibrated Bilstein shocks, a lower ride height, special alignment settings, and stouter anti-roll bars at both ends of the car.

Just like the energized engine, the modified chassis has its price. The ride is firm over any road surface, road noise is high, and lane-divider dots come through loud and clear. There's also a lot of kickback in the steering, as well as a distinct wandering on undulating roads, perhaps a result of the low ride height interacting unfavorably with the steering geometry. The low height also collaborates with the wonderfully aggressive Ruf front spoiler to make the pilot fearful of driveway ramps and curbs.

Despite its comfort compromises, the Ruf car does have its luxuries. Nearly every interior surface is covered with leather, the driver is coddled by Recaro seats and a thick-rimmed steering wheel, and the car is equipped with every factory option and an elaborate sound system. And since Ruf is small and eager to please, it's willing to provide virtually any desired interior or exterior finish.

Ruf Automobile of America (213–498–8440) will sell you a complete Turbo for $75,000. Alternatively, you can buy the various Ruf components to upgrade your own 911. Before plunking down your cash, though, make sure you're as serious about your driving as Ruf is about speed.

—*Csaba Csere*

**Vehicle type:** rear-engine, rear-wheel-drive, 2 + 2-passenger, 2-door coupe

**Price as tested:** $75,000 (base price: $75,000)

Engine type: turbocharged and intercooled flat 6, aluminum block and heads, Bosch K-Jetronic fuel injection

| | |
|---|---|
| Displacement | 205 cu in, 3367cc |
| Power (SAE net) | 360 bhp @ 6000 rpm |
| Transmission | 5-speed |
| Wheelbase | 89.4 in |
| Length | 168.9 in |
| Curb weight | 3030 lb |
| Zero to 60 mph | 4.8 sec |
| Zero to 100 mph | 10.9 sec |
| Zero to 130 mph | 22.0 sec |
| Standing ¼-mile | 13.2 sec @ 107 mph |
| Top speed | 169 mph |
| Braking, 70–0 mph | 180 ft |
| Roadholding, 300-ft-dia skidpad | 0.87 g |
| *C/D* observed fuel economy | 12 mpg |

# Porsche 911 Turbo

*The vexing return of the car
no one could forget.*

• Set your time control for 1979. Forget everything automotive you've experienced in the last six years. Let yourself drift back, back, all the way back to a time when one high-performance automobile in America stood head and shoulders above the rest.

Six years ago, things looked grim for car enthusiasts. The feds' emissions standards and a pair of fuel crises had just about squeezed the life out of hot cars—with one notable exception. Towering Colossus-like above the sea of gas-sipping econo-boxes and throbbing diesels was the Porsche 930 Turbo. Its sheetmetal bulged like Arnold Schwarzenegger's chest. Its engine had turbocharged lungs. It accelerated as if there were a Saturn booster strapped to its tail. It became the altar at which car nuts worshiped, and no one with even a few drops of 30-weight in his veins would ever forget it.

The 930 Turbo was the promise of a better tomorrow through turbocharging. But at the end of the 1979 model year, it was withdrawn from the U.S. market. The expense and the complexity of maintaining its power level while bringing its air-cooled engine into line with tightening emissions regulations were cited as the primary cause of its demise. Its penchant for gasoline (it delivered only 12 mpg on the EPA city test), its high price, and its low sales volume were the nails in the coffin. America would have to get by with normally aspirated 911s, or none at all.

This was not easy news to take. Sure, the Turbo was beyond the reach of all but a few wealthy buyers. Its passing shouldn't have meant a thing to the rest of us, but it did. That's because the Porsche 930 Turbo transcended the realm of everyday cars and parts and suggested retail prices. It de-



**PHOTOGRAPHY BY RICHARD GEORGE**

fined and dominated an era in automotive history.

It was inevitable that a car as coveted as the 930 would continue to find its way here through the gray market. It never went out of production, so a ready supply has been available for those with fat wallets; we tested a number of such cars ourselves. To thwart the gray-market traffic, Porsche went so far as to offer the 930's voluptuous bodywork and revised chassis pieces as a big-buck option on the 911 Carrera.

As of this moment, all of these substitutes for the real thing are hereby declared obsolete. Porsche Cars North America is once again importing the most potent member of its rear-engined family, this time under the 911 Turbo name.

· The manufacturer's reasons for its change of heart are simple and straightforward. Porsche has finally recognized the full importance of the North American market, where more than half of its cars are sold. As a result, we will no longer be denied the best stuff, which has been heretofore reserved for Europe. The game plan is for Porsche to offer all of its model lines here, while making every attempt to equalize power levels worldwide. Last year, we were granted the four-valve-per-cylinder 928 *before* the German market got it. The 944 Turbo makes the same power wherever it's sold. The 911 Turbo is the third step in that direction.

Importing the 911 Turbo is also the best way for Porsche to blunt the gray market and to channel the profits from U.S. sales into its own coffers. Why buy a privately federalized European-spec 911 Turbo, which might be hard to get parts for, when you 'can have a factory-fresh, EPA-approved model with the full dealer warranty?

Corporate maneuvering aside, the best part of the deal is that a solid-gold, heart-thumping supercar has returned to our midst. It's as if Ferrari had brought back the Daytona, or Ford had resurrected the Cobra. But is all the lore surrounding the mythical 930 Turbo grounded in reality, or have our warm memories been clouded by time and distance? Is the new 911 Turbo still the King Kong super ride of our demented dreams, or has automotive science passed it by? Only a test drive will tell.

To look at the new 911 Turbo is to stare right back into 1979. Only the keenest eye will notice that the rear tires now fill out the massive flared fenders a little more fully. The engineers have attacked the 930's nasty tendency to wag its tail during hard cornering by specifying wider-than-ever, 245/45VR-16 Dunlop SP Sport D40 rear tires in place of the old car's 225/50VR-16 rubber. The bigger tires are mounted on 9.0-inch wheels, which are an inch wider than before.

Precious few cars could live through six years without so much as a face lift, but the 911 Turbo has done just that. This car has a sexier body than Madonna, and the years have dulled its charm not a whit. We sampled the 911 Turbo in the L.A. area, which has the highest per-capita number of winged and flared 911s in the Western Hemisphere, but our red beast wowed the masses nonetheless. They still find this a spellbinding automobile, and far more folks than you'd expect went out of their way to let us know that.

Inside, the Turbo could be any 911 of recent vintage, but for a few minor details. A small boost gauge is incorporated into the tachometer at the six-o'clock position. Check the standard plastic shift knob and you'll see that the gear pattern stops at fourth. (Turbos have never been equipped with five-speeds by the factory.)

Aside from that, the Turbo is just a well-dressed 911. Soft, sweet-smelling leather is lavished on the cockpit, including the dash top. A full load of extras, from air to sunroof, are standard—just as you'd expect in a car that comes in at a nice, round $48,000. But that's it. No surprises or great advances have sprung up since we last saw this model.

You won't find any major changes under the whale tail, either. The 911 Turbo's air-cooled flat six is basically the same one that tantalized us so much back in 1979. The turbocharged and intercooled powerplant still displaces 3.3 liters, and such details as its bore, stroke, and compression ratio remain unchanged.

The bottom line—the horsepower coming off the end of the crankshaft—is fatter than ever, though. Porsche's data banks are six years richer with emissions-control knowledge, and it's been put to good use in the 911 Turbo. The tweaking includes a three-way catalytic converter, an oxygen sensor, and electronic assistance for the Bosch mechanical fuel-injection system. In 1979, Porsche was carping about the difficulties of making its air-cooled powerplant comply with federal exhaust-emissions standards. Today, the engineers have made it comply, adding an impressive 29 hp in the bargain. They also managed to improve fuel economy by 33 percent, though the 911 Turbo's 16 mpg still isn't good enough to get it past the gas-guzzler law. This scrape with the tax man adds a $500 penalty to the car's base price.

Nevertheless, if 1979 was a great year for turbocharged 911s, 1986 ought to be even better, right? Twenty-nine more horses, fatter tires, and six years of chassis development could only make things positively dreamy.

There's certainly no shortage of promise when you get the proceedings under way. On a cool morning, the beat of the 911 Turbo's idle will warm you faster than the heater. This engine sounds serious: lumpy and hoarse, with an occasional *spit!* thrown in for good measure.

There's no need to hound the 911 around town. Enough torque is on hand

the test track again. There was certainly nothing wrong this time around. Big horsepower, big rear tires, and a big rear weight bias enabled our second test car to blow out of the hole like a cannon shot. With a searing 0-to-60-mph run of 4.6 seconds and a clocking of 13.1 seconds at 105 mph through the quarter-mile, the 911 Turbo is most assuredly this season's acceleration ace—providing you're willing to resort to rough, wheel-spinning, drag-race starts.

Out on the road, though, these numbers pale next to the Turbo's boost-lag arthritis. Even the healthier of our two test cars took forever to spin its turbo up to liftoff speed. Once it was up and running, it was plenty strong, but it just didn't awe us the way the old 930 used to.

Then again, there's more to our memories of the 930 than pure speed. It was also known as one of the trickiest handlers around. Driving one hard was a job for experts. Putting the power on aggressively in a corner would pitch the nose way up, and the 930 would try to run straight over its front tires. Lift off the gas just a few millimeters in these conditions and the 930 would swing sideways so fast, it would jump-start your heart.

Not so the new 911 Turbo. On the tortured curves of California's Ortega Highway, it shows real poise. In the last six years it's obviously been taught some manners. Antics that would have spun you out before hardly faze it. The brakes are superb. It's still hard work to drive very, *very* fast, but it's much more forgiving now.

Comparing this experience with our last 930 outing, in 1979, it's clear that things have changed. The 930 was deadly in the curves and awesome on the straights, and the 911 Turbo is mellower in both areas.

This pass through the time barrier, the 911 Turbo's performance just hasn't blown our minds—and we think we know why. Back in 1979, there really wasn't any other car in America that offered anywhere near the 930's kind of speed. Today, however, we're in the middle of a horsepower boom. We've got 157-mph 944 Turbos, 154-mph 928s, 151-mph Corvettes, 140-mph Camaros—hell, even Saab is in the 140-plus club these days.

Faced with these facts, we can draw no other conclusion than that the handwriting is on the wall for the 911 Turbo. Precious few cars can sprint with it, but the march of technology has produced a whole flock of turbo cars with much better manners. This is, no doubt, why Porsche is hard at work on a four-valve-per-cylinder version of this car, and why the awesome 959 prototype is fitted with a sophisticated twin-turbocharger setup.

But this is today—the here and now. Taking a cold, hard look at the 911 Turbo's vexing return, we get the feeling that fond memory may have been better left undisturbed. —*Rich Ceppos*

for easing along in thick L.A. traffic without fishing for boost. But look out the first time you decide to scoot away from a light. First gear is as steep as the north face of the Eiger—it's good for 50 mph—and there's no heavy thrust down low. A cheerleader in a clapped-out Mustang II will have no trouble beating you across an intersection while checking her makeup. As a matter of fact, one did exactly that to us.

Then the boost comes in as the revs go past 4500 rpm, the exhaust hisses like a very angry 3000-pound cat, and *whoosh!* you rattle the Mustang's windows as you blow by.

On the freeway, locked in a clot of 65-mph traffic, the Turbo feels dead on its feet. Rolling along in fourth gear with the engine just ticking over, it's a good five-count before the boost needle moves off of the peg. Drop down to third and it's still a three-count before the rockets fire and you can blast through a hole into the next lane.

This is no fun. Your average Volvo 740

Turbo would be ten car-lengths down the passing lane by now. In truth, second gear, which goes all the way to 86 mph, is the way to deal with the freeway—but it's kind of embarrassing, not to mention noisy, howling along at 4500 rpm just to have the horsepower on retainer.

We remember the 930 as having bags full of boost lag, but was it really this bad? Has turbo technology left this car—a *Porsche*—so hopelessly behind?

Our track testing indicated that something was probably wrong with our test car. Its clutch was definitely slipping, and we suspect that a waste-gate problem kept the engine from building boost quickly. This car also suffered a thrown A/C drive belt and a recalcitrant driver's door during our testing, so it was not the best example of Porsche quality we've seen.

Further study was called for, so we traded our flaming-red 911 Turbo for a deep-blue-metallic number (yes, you do see two different cars in the photos) and set off for

**Vehicle type:** rear-engine, rear-wheel-drive, 2+2-passenger, 2-door coupe

**Price as tested:** $49,720

**Options on test car:** base Porsche 911 Turbo, $48,000; limited-slip differential, $595; extended steering-wheel hub, $47; gas-guzzler tax, $500; freight, $578

**Standard accessories:** power windows, seats, locks, and sunroof, A/C, rear defroster and wiper

**Sound system:** Blaupunkt AM/FM-stereo radio/cassette, 4 speakers

### ENGINE

| | |
|---|---|
| Type | turbocharged and intercooled flat 6, aluminum block and heads |
| Bore x stroke | 3.82 x 2.93 in, 97.0 x 74.4mm |
| Displacement | 201 cu in, 3299cc |
| Compression ratio | 7.0:1 |
| Fuel system | Bosch KE-Jetronic fuel injection |
| Turbocharger | KKK |
| Waste gate | Porsche |
| Maximum boost pressure | 10.2 psi |
| Power (SAE net) | 282 bhp @ 5500 rpm |
| Torque (SAE net) | 278 lb-ft @ 4000 rpm |

### DRIVETRAIN

Transmission . . . . . . . . . . . . . . . . . . . . . . . . . . 4-speed
Final-drive ratio . . . . . . . . . . . . . . . 4.22:1, limited slip

| Gear | Ratio | Mph/1000 rpm | Max. test speed |
|---|---|---|---|
| I | 2.25 | 7.5 | 50 mph (6700 rpm) |
| II | 1.30 | 12.9 | 86 mph (6700 rpm) |
| III | 0.89 | 18.9 | 127 mph (6700 rpm) |
| IV | 0.63 | 27.0 | 155 mph (5750 rpm) |

### DIMENSIONS AND CAPACITIES

Wheelbase . . . . . . . . . . . . . . . . . . . . . . . . . . . . . . 89.4 in
Track, F/R . . . . . . . . . . . . . . . . . . . . . . . . . . 56.4/58.7 in
Length . . . . . . . . . . . . . . . . . . . . . . . . . . . . . . 168.9 in
Width . . . . . . . . . . . . . . . . . . . . . . . . . . . . . . . 69.9 in
Height . . . . . . . . . . . . . . . . . . . . . . . . . . . . . . . 51.6 in
Frontal area . . . . . . . . . . . . . . . . . . . . . . . . 20.0 sq ft
Curb weight . . . . . . . . . . . . . . . . . . . . . . . . . . 3040 lb
Weight distribution, F/R . . . . . . . . . . . . . 37.5/62.5%
Fuel capacity . . . . . . . . . . . . . . . . . . . . . . . . 22.5 gal

### CHASSIS/BODY

Type . . . . . . . . . . . . . . . . . . . . . . . . . unit construction
Body material . . . . . . . . . . . . . . . welded steel stampings

### INTERIOR

| | |
|---|---|
| SAE volume, front seat | 43 cu ft |
| rear seat | 13 cu ft |
| trunk space | 5 cu ft |
| Front seats | bucket |
| Seat adjustments | fore and aft, seatback angle, front height, rear height |
| General comfort | poor fair good **excellent** |
| Fore-and-aft support | poor fair good **excellent** |
| Lateral support | poor fair **good** excellent |

### SUSPENSION

F: . . . . . . . . ind, strut located by a control arm, torsion bars, anti-roll bar
R: . . . . . . . . . ind, semi-trailing arm, torsion bars, anti-roll bar

### STEERING

Type . . . . . . . . . . . . . . . . . . . . . . . . . rack-and-pinion
Turning circle curb-to-curb . . . . . . . . . . . . . . . . 35.9 ft

### BRAKES

F: . . . . . . . . . . . . . . . . . . . . . . 12.0 x 1.3-in vented disc
R: . . . . . . . . . . . . . . . . . . . . . . 12.0 x 1.1-in vented disc

### WHEELS AND TIRES

Wheel size . . . . . . . . . . . . . . . F: 7.0 x 16 in; R: 9.0 x 16 in
Tires . . . . . . . . . . . Dunlop SP Sport D40, F: 205/55VR-16; R: 245/45VR-16

## CAR AND DRIVER TEST RESULTS

### ACCELERATION

| | Seconds |
|---|---|
| Zero to 30 mph | 1.6 |
| 40 mph | 2.3 |
| 50 mph | 3.3 |
| 60 mph | 4.6 |
| 70 mph | 5.8 |
| 80 mph | 7.3 |
| 90 mph | 9.9 |
| 100 mph | 11.9 |
| 110 mph | 14.7 |
| 120 mph | 18.9 |
| Top-gear passing time, 30–50 mph | 11.1 |
| 50–70 mph | 11.4 |
| Standing ¼-mile | 13.1 sec @ 105 mph |
| Top speed | 155 mph |

### BRAKING

70–0 mph @ impending lockup . . . . . . . . . . . . 173 ft
Modulation . . . . . . . . . . . . . . . poor fair **good** excellent
Fade . . . . . . . . . . . . . . . . . . . . **none** moderate heavy

Front-rear balance . . . . . . . . . . . . . . . . . poor fair **good**

### HANDLING

Roadholding, 300-ft-dia skidpad . . . . . . . . . . . 0.82 g
Understeer . . . . . . . . . . . . **minimal** moderate excessive

### COAST-DOWN MEASUREMENTS

| Road horsepower @ 30 mph | 6 hp |
|---|---|
| 50 mph | 14 hp |
| 70 mph | 30 hp |

### FUEL ECONOMY

EPA city driving . . . . . . . . . . . . . . . . . . . . . . 16 mpg
EPA highway driving . . . . . . . . . . . . . . . . . . 22 mpg
C/D observed fuel economy . . . . . . . . . . . . . **13 mpg**

### INTERIOR SOUND LEVEL

| Idle | 64 dBA |
|---|---|
| Full-throttle acceleration | 83 dBA |
| 70-mph cruising | 77 dBA |
| 70-mph coasting | 74 dBA |

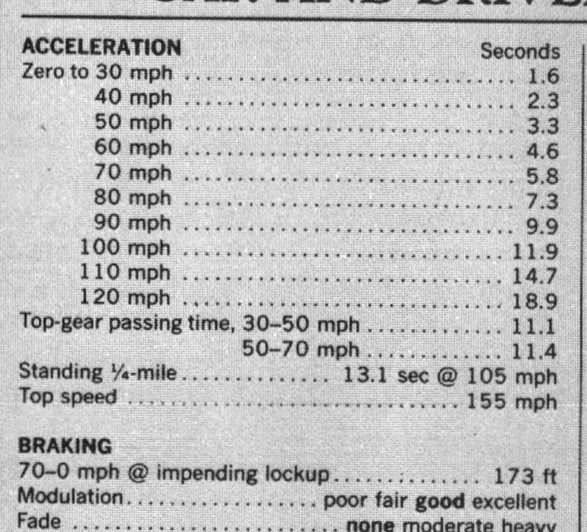

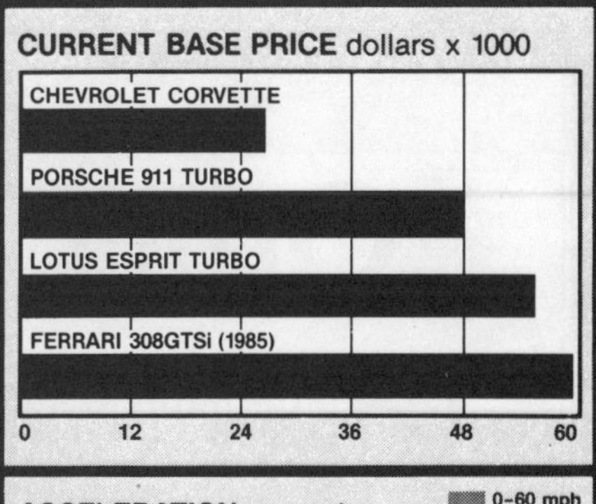


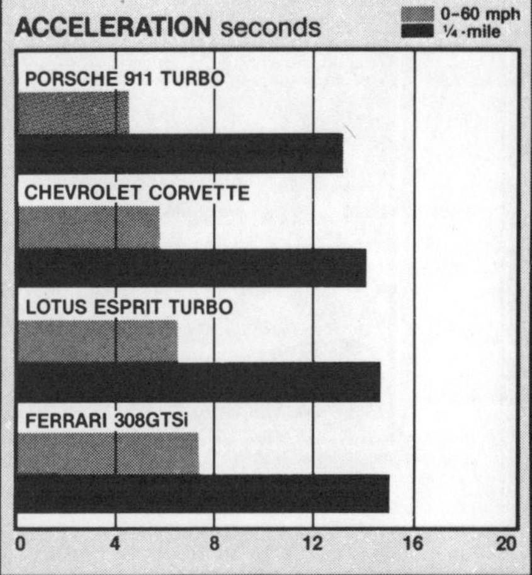


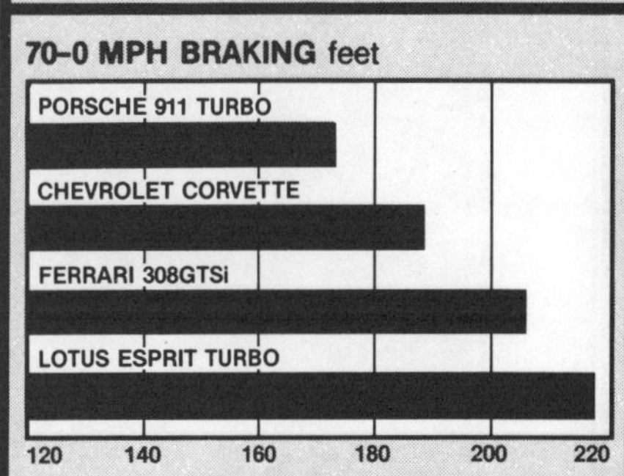


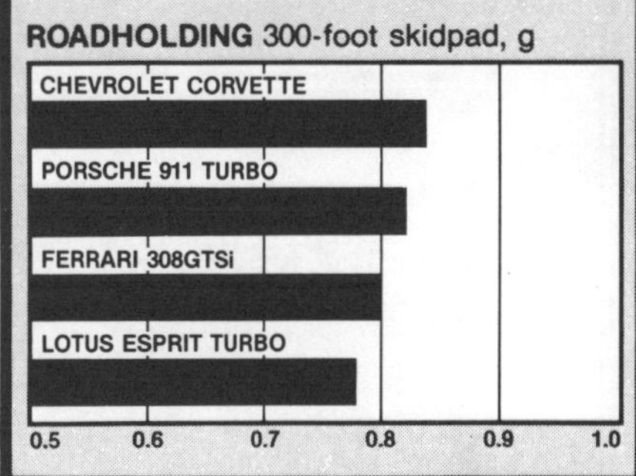


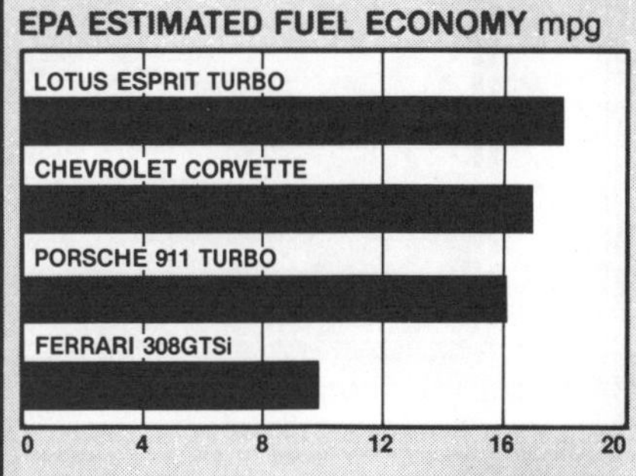

# CAR AND DRIVER

# Other Titles in this Series

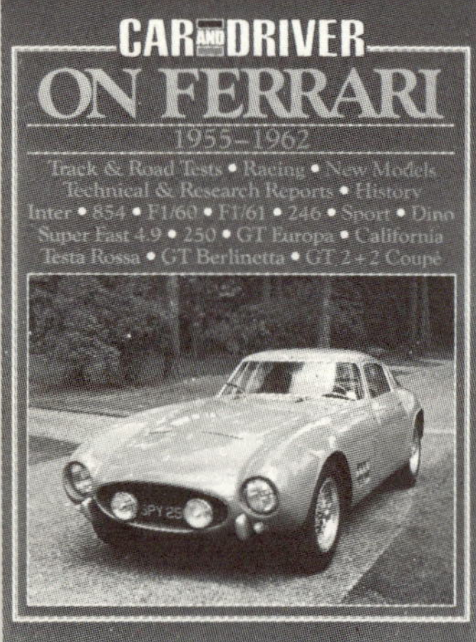

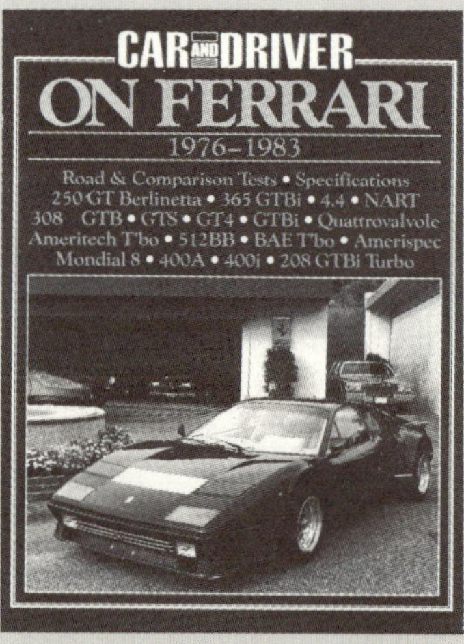

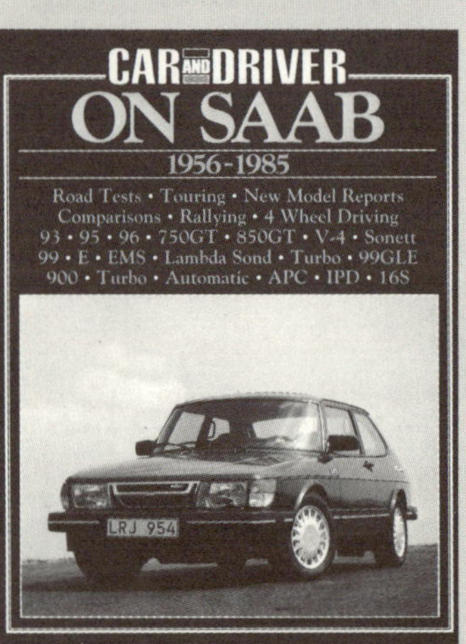